FIGHTING
IN THE
DARK

FIGHTING
IN THE
DARK

NAVAL COMBAT AT NIGHT, 1904–1944

EDITED BY

VINCENT P. O'HARA AND TRENT HONE

NAVAL INSTITUTE PRESS
ANNAPOLIS, MARYLAND

Naval Institute Press
291 Wood Road
Annapolis, MD 21402

Library of Congress Cataloging-in-Publication Data
Names: O'Hara, Vincent P., editor. | Hone, Trent, editor.
Title: Fighting in the dark : naval combat at night: 1904–1944 / edited by
  Vincent P. O'Hara, and Trent Hone.
Identifiers: LCCN 2022043022 (print) | LCCN 2022043023 (ebook)
  | ISBN 9781682477809 (hardcover) | ISBN 9781682477816 (ebook)
Subjects: LCSH: Naval history, Modern—20th century. | Night fighting
  (Military science) | Navies—History—20th century. | BISAC: HISTORY /
  Military / Naval | TECHNOLOGY & ENGINEERING / Military Science
Classification: LCC D436 .F54 2023 (print) | LCC D436 (ebook) |
  DDC 359.00904—dc23/eng/20230103
LC record available at https://lccn.loc.gov/2022043022
LC ebook record available at https://lccn.loc.gov/2022043023

31 30 29 28 27 26 25 24 23   9 8 7 6 5 4 3 2 1
First printing

Maps in chapter 5 created by John Parshall; maps in chapter 7 adapted from originals by Chris Johnson and Robin Brass. All other maps created by Vincent O'Hara.

CONTENTS

ILLUSTRATIONS

FIGURES

PHOTOGRAPHS

PREFACE

Enemy warships interrupted HMAS *Perth*'s night transit of Sunda Strait. One crewman "saw enlarging spots of light on two Japanese destroyers and knew they were opening the shutters of their searchlights. Then the 4-inch cracked and put the lights out . . . . something crashed against the gun shield close to his head and spun into the deck at his feet. A star shell flowered, high and brilliantly soft, and he looked down and saw a chunk of jagged metal, about six inches long, impaled in the deck . . . everything was, for a long moment, as hushed as the bush at noon, before a great pillar of water and oil collapsed on him. He saw a [dim shape] and from [it] was pouring stream after stream of red and blue and amber tracer as though madmen were throwing electric light bulbs across the sky."[1]

This staccato account communicates danger, confusion, and a sense of alien beauty. It serves to fix the image of this book's subject: naval combat at night and the challenges it imposed as officers and sailors learned to fight in an unfriendly environment and win success using new platforms, tools, and the innovative tactics that came with them.

Conflict is a constant of human experience, yet few organizations manage conflict well, particularly when change invalidates established practices and presents new challenges, new dangers, and new opportunities. Despite their specialization in conflict, militaries are especially susceptible to this. In part, this is because the requirements and economics of peace are different from those of war, complicating the incorporation of new tactics and technologies during peacetime. And what is one result of this dynamic? That surprise is "the most consistent element on the battlefield." The military that can most quickly adapt under wartime pressure will enhance its chances of victory.[2]

Historically, this has been particularly true for navies. Warships are expensive. Warships require sophisticated and specialized equipment. Warships take years to design and build, and every ship design embeds assumptions about how best to win naval conflicts. Generally, these are based on past experiences but assumptions about untested technologies can be overly optimistic or erroneous. Because of their cost, relatively few ships can be built, so a seemingly minor advantage can give a substantial edge. Mistakes linger.

In the first half of the twentieth century, naval warfare underwent a period of rapid evolution. In 1900 naval combat was conducted on the surface of the ocean and almost exclusively by day. By 1945 naval warfare was conducted on the ocean's surface, beneath the surface, and in the air as well. New weapons and tools had given navies the ability to see in the dark, to communicate across thousands of miles, and to strike at targets two hundred miles away instead of two. Eighty percent of surface combat occurred at night. The reasons that night combat came to dominate are many and complex.

There were many reasons that navies avoided night combat. It is hard to hit a practically invisible, moving target with aimed gunfire. Up through the nineteenth century, ships were deaf and mute at night beyond the sight of a lantern or the sound of a cannon shot. Prior to the nineteenth century, only one sea battle out of ten was nocturnal, and most of those occurred as the result of accidental encounters or as continuations of an action that began in daylight.

The process by which navies learned to use the dark and adapted to it as a medium for effective combat was difficult—more so for some navies than for others. This book is about that process. The platforms, such as small torpedo boats; the weapons, such as torpedoes; and then rapid-fire guns and new tools, such as searchlights and radio—are all important because they made night combat effective. But how these platforms and weapons were used; how they were integrated into a coherent system; and how training, doctrine, and even strategy were affected are an equally crucial part of the story. Finally, the major navies of the world took different paths to integrate night combat into their plans and tactics. The nature of these variations and the reasons for them are an important part of the tale.

Fighting in the Dark examines seven periods from 1904 to 1944, focusing on a particular navy in each period:

- ▸ The Russian navy in 1904–1905
- ▸ The German navy in 1914–1916
- ▸ The British navy in 1916–1939
- ▸ The Italian navy in 1940–1943
- ▸ The Japanese navy in 1922–1942
- ▸ The U.S. Navy in 1942–1944
- ▸ The British and Canadian navies in 1943–1944

The chapters that follow consider how these navies confronted the specific challenges of night combat between surface warships and how they mastered new skills and technologies. It is a story of discovery and practice, innovation and invention. It is a story of persistence and growing competence, although not in all cases. Expertise was a product of training, technology, and a willingness to *embrace* risk—not just accept it—to gain significant advantage.

This work uses certain conventions. Unless otherwise stated, miles are nautical miles (nm) and times are local. Times are expressed in twenty-four hour format: 8:10 a.m. is 0810 and 8:10 p.m. is 2010. In respect to naval tradition, and consistent with the conventions of the time, ships will be referred to as "she" and "her." The editors and authors have agreed, after some debate, to exclude night combat involving submarines or coastal forces—not only for reasons of length, but also in the hope that focusing on surface vessels that fought similar types of actions during the period might help highlight the differences in how night combat developed in each of the world's navies. We have also agreed to regard as a "night action" a battle that occurs all or in part when the sun is below the horizon. Visibility was a crucial factor. There were many daylight battles—the March 1942 Battle of Sirte immediately comes to mind—where stormy weather compromised visibility to a greater degree than some night actions, but not in all cases and in all ways. Those actions are not considered here.

This book is about naval combat at night.

ACKNOWLEDGMENTS

The editors are grateful to the authors for agreeing to participate in this project and for the enthusiasm and deep knowledge they brought to the table. Their talent, expertise, and professionalism show in the pages that follow. Thank you Steve, Len, James, Enrico, Jon, and Michael. We also thank Glenn Griffith and Adam Kane at the Naval Institute Press, who are a joy to work with and quickly perceived the value of our concept. O'Hara further thanks Maria, Yunuen, and Vincent, for their unfailing support, without which he feels his work would be impossible. Hone further thanks Stacie Parillo and Elizabeth Delmage of the Naval War College Archives and Barry Zerby, Rick Peuser, Mark Mollan, and Charles W. Johnson of the National Archives for their invaluable assistance locating material used in the manuscript. Hone is most grateful for the assistance of his wife, Lauren Hillman, who made his participation in this project possible.

ABBREVIATIONS

AA	antiaircraft
AIC	Action Information Center
AID	Action Information Intercom
AIO	Action Information Organization
APG	*Apparecchio di Punteria Generale* (general night aiming device)
AR	Action Report
ARL	Admiralty Research Laboratory
ASV	Radar, Air-to-Surface Vessel
BatDiv	Battleship Division
BCE	Before the Common Era
CIC	Combat Information Center
C-in-C	Commander in Chief
CMB	coastal motor boat
CTF	Commander, Task Force
CTG	Commander, Task Group
DCNS	Deputy Chief of the Naval Staff
DF	direction-finding or destroyer flotilla
DHH	Directorate of History and Heritage
DP	dual-purpose
DTSD	Directorate of Tactical and Staff Duties
FC	U.S. Navy Mark 3 Fire-Control Radar
FH	U.S. Navy Mark 8 Fire-Control Radar
G.G.	*Grande Gittata*
GFBO	Grand Fleet Battle Orders
GRT	gross registered tonnage
HF	high-frequency
HMAS	His/Her Majesty's Australian Ship
HMCS	His/Her Majesty's Canadian Ship

HMNZS   His/Her Majesty's New Zealand Ship

HMS   His/Her Majesty's Ship (British)

IFF   identification friend or foe

IGM   Instituto di Guerra Marittima (Naval War Institute)

IJN   Imperial Japanese Navy

LCS   Light Cruiser Squadron

LOP   letter of proceedings

MAS   Motoscafo Armato Silurante (torpedo-armed motorboat)

MS   Motosiluranti (large MTB)

MTB   motor torpedo boat

NARA   National Archives of the United States

ONI   U.S. Navy Office of Naval Intelligence

ORP   Okręt Rzeczypospolitej Polskiej (Warship of the Republic of Poland)

OTC   Officer in Tactical Command

Pdr   pounder (denotes shell weight)

PPI   plan position indicator

RAF   Royal Air Force

RCN   Royal Canadian Navy

RN   Royal Navy (British)

SG   type of U.S. centimetric radar

SIGINT signals intelligence

TBS   VHF voice radio

TNA   The National Archives of the United Kingdom

TNT   trinitrotoluene

TPA   Telefono per Ammiragli

USN   United States Navy

USS   United States Ship

VHF   very high frequency

W/T   wireless telegraphy (radio)

INTRODUCTION

From the time Ramesses III defeated the Sea Peoples in the twelfth century BCE, and undoubtedly before, there have been naval battles. The vast majority of these have been fought in daylight. Why? In combat, whether on land or sea, a commander's greatest challenge is to maintain the cohesion of his or her forces, so that they work toward a common goal. That can be done in many ways—through inspiring personal leadership, well-managed communications, and coherent tactics and doctrine, to name just three. But to lead effectively, a commander must understand what is happening and provide subordinates with effective direction. In the dark it is much harder to grasp a situation and issue clear instructions. On land any night action—no matter how ready the troops, how good the preparations, and how great the general—will quickly illustrate Carl von Clausewitz's concepts of "fog" and "friction." Frustration and confusion will emerge and even "the simplest thing [will become] ... difficult."[1]

At sea the challenges are vastly greater. One reason is that at night visibility on the sea is extremely variable. A moonless foggy night in the North Sea and a clear moonless night in the Pacific are completely different environments. The moon can dramatically enhance visibility or reduce it, depending upon its phase, elevation, and position. Weather effects, such as clouds, rain, fog, and sea state, compound the uncertainty. As ships steam through squalls and clouds wend their way across the moon, visibility can change from one second to the next.

Such factors disrupt the sense of distance and scale. The human eye can adjust to darkness, but it can take fifteen to thirty minutes to reach peak performance. And even then, the mind filters what the eye sees, based on expectations and training. Officers and sailors tend to "observe" what they anticipate. The history of naval combat at night is replete with misidentifications and erroneous reports—rain showers mistaken for nearby land; enemy ships reported as friendlies; destroyers taken for cruisers; hits "scored" when all shots missed; and ships "sunk," when in fact the target was a rocky islet or even the radar signature of flocking birds.[2] For those who have navigated the night sea's amorphous darkness, such errors are ordinary. In the dark, the eye plays tricks on the mind, and the mind is all too eager to return the favor.

As a result, large naval forces avoided night combat for much of history. Instead, smaller forces used darkness to gain advantage, leveraging stealth and surprise to seize the initiative and disrupt an opponent's cohesion. The English fireship attack on the Spanish fleet in 1588 is an excellent example. Night combat was always a gamble, but under the right circumstances—and with the right tools and weapons—well-trained captains and crews could make it a profitable one.

BEAR UP AND SAIL LARGE: NIGHT COMBAT IN THE AGE OF SAIL

This principle applied throughout the age of sail. In the fighting instructions issued by the (British) Navy Royal in 1653, the only provision for dealing with night combat specified "[t]hat if any engagement by day shall continue till night and the general shall please to anchor, then upon signal given they all anchor in as good order as may be . . . and if the general please to retreat without anchoring, the signal to be firing two guns . . . and within three minutes after to do the like with two guns more."[3] However, night combat could be required under "the exigencies of his master's service," and preparations to "attack or repulse [an attack] by night" grew more detailed to address the challenges of nocturnal command and control.[4] By the late eighteenth century, the Royal Navy's instructions for night signaling and combat had reached a level of detail that would continue until technological changes in the late nineteenth century required further revision and expansion.[5]

In the Royal Navy the preferred formation for night combat was line-of-battle. This facilitated control because each ship could follow the one ahead. Signals were communicated by hanging lanterns in specific configurations rather than using signal flags. For example, to indicate an upcoming tack, ships were instructed to "show two lights, one under the other, at the bowsprit end."[6]

Still, the few night fleet engagements that did occur were often the continuation of an action that had begun during the day. The Battle of Cape St.

Vincent, fought on 16–17 January 1780, commenced two hours before dark, when eighteen British line-of-battle ships, supported by six frigates, chased down a Spanish force half as large. It continued long into a bright moonlit night. Another example is Admiral Horatio Nelson's victory at the Battle of the Nile, which raged through the night of 1–2 August 1798. In the Aegean, a Venetian fleet tangled with the Ottomans from late afternoon into the dark on 12 June 1717. There was a larger repeat on 19 July 1717 when, after a long day action, twenty-six Venetian and four Portuguese ships of the line fought twenty-six Ottoman capital ships well into the night.[7]

A list of all minor surface engagements fought by the Royal Navy from 1741 through 1814 indicates that 20 of 207 actions, or 9.7 percent, were fought at night. Thus, in the period before the nineteenth-century technological revolution, roughly 1 naval action in 10—large and small—was fought at night. During World War II, between 1944 and 1945, warships larger than 500 tons' displacement fought 45 naval actions, and 36 of those, or 80 percent, were nocturnal. Many factors drove this change, but technology triggered it.

ENGINES OF CHANGE

Naval technology improved gradually during the six centuries prior to the mid-1800s. In 1850 a ship of the line was far larger and more powerful than one of 1650, but it was essentially the same platform. A crew from 1650 would have been able to work and fight an 1850 vessel with minimal familiarization. The British navy's fighting instructions for 1653 applied, in large part, equally well two hundred years later. In the middle of the nineteenth century, however, a surge of new technology brought on by the industrial revolution swept the globe. Men born during the French Revolution found themselves navigating a new world that not even the fantastic imagination of Jules Verne had conceived.

From Sail to Steam

First among the great changes that revolutionized naval warfare was the development of steam locomotion. This freed ships to go wherever they wished, at higher speeds than wind power, but at a cost. Warships had to carry their own energy with them, in the form of coal. This reduced their operational radius; three or four days of high-speed steaming would empty their coal bunkers. As a form of auxiliary locomotion, major steam warships retained sailing rigs for half a century. Larger navies developed networks of coaling stations to keep their ships supplied and secure the trade routes that were vital to their interests.

Along with steam, nineteenth-century advances in metallurgy revolutionized weaponry and defenses against it. Increasingly sophisticated armor plating protected warships. More powerful breech-loading guns and specialized projectiles challenged that protection. Better propellants, such as cordite, gave those guns greater muzzle velocities, range, and accuracy. More powerful explosives, like guncotton and TNT, made the shells they filled ever more deadly.

The emergence of steam propulsion and better weapons forced naval officers to devise new ways to fight. The half-century in which these developments occurred most intensely, 1850–1904, was a period of relative peace. There were no significant naval campaigns like those of the seventeenth and eighteenth centuries. Few nations fought major naval engagements, and technology quickly outpaced the lessons they provided. For example, the 1866 Battle of Lissa between Austria and Italy suggested that the ram had returned as an important weapon of war. For several decades, major warships were designed with ram bows, but higher speeds and longer-range guns rapidly made such arrangements obsolete. The ram bow's most infamous "success" was the sinking of battleship HMS *Victoria* by HMS *Camperdown* during maneuvers in 1893.

Steam, armor, and more powerful guns forced navies to adjust their tactics and plans, yet there was little certainty about how to do so. A deep well of collective experience, from which every navy drew, was completely drained within a generation. Naval leaders such as British Admiral Charles Hotham, who served his apprenticeship on the wooden-screw frigate HMS *Curacoa* and later commanded the fleet at the 1902 Royal Review from the bridge of HMS *Royal Sovereign*—a 14,000-ton battleship armed with two pairs of 13.5-inch breech-loading guns—were forced to rely on theories, experimentation, and intelligent guesswork to determine how best to use their forces.[8] Given the rapid changes that characterized this period, tactical and technical uncertainty was pervasive. This uncertainty was compounded by the introduction of several new technologies that incentivized fighting at night, foremost among them the torpedo, an innovative combination of explosive, motor, and inertial guidance system.

Torpedoes

When the first locomotive torpedo emerged from the Austro-Hungarian workshop of Englishman Robert Whitehead, it threatened to upset the existing principles of naval power. Torpedoes could be carried on a wide variety of ships, including small, "expendable" platforms. Since they delivered their explosives underwater, torpedoes threatened even the largest, most well-armored ships. In the words of the British Admiralty, "the most powerful ship is liable to be destroyed by a torpedo projected from a vessel of the utmost comparative insignificance."[9] Before only a battleship could sink another battleship; now many ships could do so, challenging the notion that a battle fleet was essential to exert sea control.

The torpedo was the perfect weapon for smaller forces that sought to use night battle to gain an advantage. In addition to being a potential battleship-killer, torpedoes were inherently stealthy, especially at night. Darkness could cover the approach of a small force, conceal it while it launched its weapons, and then hide its withdrawal. Because torpedoes were "fire and forget" weapons, there was no need to close or continue to engage once they were away. However, at the dawn of the torpedo era, these tactics were theoretical, and combat quickly revealed the torpedo's limitations. In 1877 the steam frigate HMS *Shah* first fired a torpedo in anger against the Peruvian turret ship *Huáscar*. The target spotted the torpedo, reversed course, and, steaming at eleven knots, outran it. Through 1894 torpedoes were used on several occasions during the Russo-Turkish war (1877–78) and the War of the Pacific (1879–84), but the only vessel actually sunk was an anchored Ottoman auxiliary. From these closely observed conflicts it was clear that the torpedo required greater range and speed, a more powerful warhead, and a better guidance system.

During the Chilean Civil War of 1891 two torpedo boats sank the Chilean armored frigate *Blanco Encalada* in Caldera Bay. The torpedo boats entered the bay after midnight, approached *Blanco Encalada* from opposite directions, closed to within a hundred yards, and sank her with a torpedo. This attack highlighted the potential of surprise attacks under cover of darkness. Night attacks allowed torpedo boats to make their approach in greater security; even if they were fired upon—as one of the torpedo boats was—they were more difficult to hit in the darkness and had a better chance of reaching firing range.

Continued improvements in torpedo performance over the ensuing decades increased the potential of stealthy night attacks. Large-scale adoption of gyroscopic guidance in 1898 led to a major increase in range and accuracy, further enhancing the lethality of torpedo craft and forcing fleets to revise their defenses and tactics.[10] By the time of the Russo-Japanese War, the latest torpedoes significantly outperformed their predecessors of 30 years before: ranges had nearly tripled from 800 yards to 2,200 yards, speeds had quadrupled from 6 knots to 26 knots, and warheads were a dozen times heavier, growing from 26 pounds to 325 pounds.

The Constant Quest for Advantage: New Platforms

Advancements in propulsion and naval architecture encouraged navies to develop new torpedo platforms. The Royal Navy, for example, built torpedo-armed vessels ranging in size from the *Lightning* (32.5-ton, 1876) to the 2,640-ton *Polyphemus* (1882). Heavy guns of the period fired slowly and were difficult to aim, especially in the dark. Navies quickly recognized that small, agile ships stood the best chance of closing the range and scoring a torpedo hit. They would also be the most difficult to detect at night. Eventually, most navies settled on warships with displacements of between 100 and 200 tons and called them torpedo boats.

At night, torpedo boats were expected to approach their targets with stealth. At lower speeds, the coal-burning vessels left smaller wakes and discharged fewer embers and sparks. Once the torpedo boats were within a few hundred yards of their target, they would accelerate and unleash their torpedoes.[11] The first nocturnal test of these new platforms with "modern" torpedoes came in the Sino-Japanese War of 1894–95. On 4–5 February 1895 the Japanese torpedo boat *TB-23* (85 tons, 1892) sank the Chinese capital ship *Ting Yuan* in Weihaiwei harbor. The following night, Japanese torpedo boats penetrated the harbor again and sank three warships and auxiliaries. These results suggested that night attacks by torpedo boats were "something which even the strongest fleet might be unable to avoid in the future."[12]

In the late 1880s, the French had begun building torpedo boats designed for high-seas operations; the British needed something to catch them. The resulting ships combined high speed, torpedo armament, and rapid-firing guns. As navies recognized the potential of these vessels, initially called "torpedo boat destroyers" and then simply "destroyers," they quickly increased in size and tactical utility. In the 1890s, the earliest destroyers displaced about three hundred tons. By the turn of the century, they had grown to five hundred tons.

Destroyers proved to be effective platforms for torpedo attack as well, and by the early 1920s they had assumed this duty in many of the world's navies. Torpedo boats—even the French high-seas boats—lacked true "blue-water" capability because of their small size. Destroyers were more capable. By the end of World War I, typical destroyers were more than one thousand tons' displacement; they were armed with three or four 4-inch to 5-inch guns and mounted between four and twelve 17.7-inch to 21-inch torpedoes. Destroyers had the range and seakeeping ability to operate with the fleet, serving as scouts, defensive screening vessels, and offensive strike forces. In both world wars, the destroyer was an excellent platform for night combat.

Rapid-Firing Guns

Larger ships also needed to be able to defend themselves against torpedo attack. Since it was difficult to see very far at night—the effective limit of visibility rarely exceeded eight thousand yards—guns with high rates of fire were necessary to successfully engage a torpedo boat or destroyer and disable it before it reached firing position. At the time, these were known as rapid-firing or quick-firing guns. Traditional weapons used bagged propellants and had to be sponged out after each shot to clear potentially dangerous debris from the firing chamber. Rapid-firing guns avoided this delay by using brass cartridge cases. Combined with improvements in breech mechanisms, this change significantly increased rates of fire and thereby improved the ability of large ships to defend themselves against small torpedo vessels. By the time of the Russo-Japanese War, 3-inch (12-pounder) rapid-firing guns were in service. Because even these small weapons comfortably outranged contemporary torpedoes, they created a deadly zone that torpedo craft had to transit before they could launch their weapons, thus forcing navies to develop new tactics for torpedo attack.

Searchlights

Searchlights were one of the first of the electromagnetic technologies to go to sea. The technological underpinnings that made such instruments possible were the electric arc light, an early nineteenth-century development, followed by the "electro-magnetic induction machine" (dynamo), which supplied the high current that was required to power the arc light. The searchlight's most obvious use was to illuminate the dark. It seemed an ideal way to acquire targets at night. In the British navy, the 1873 Torpedo Committee considered the electric arc light (along with pyrotechnics) as one of the best ways to defend a vessel against stealthy nocturnal torpedo attacks.[13] Accordingly, the first electric arc light to go to sea was in 1873 on board the gunboat HMS *Comet*. As one U.S. Navy officer wrote in 1882, the most likely anticipated target was enemy torpedo boats. "The electric light in a man-of-war is not to be regarded merely as an improvement or convenience, but strictly as a weapon of offense or defense. It is recognized as an invaluable defense against torpedo attacks and owes its adoption into the navies of the world to the great development of torpedo warfare."[14]

However, as noted in an 1894 article, searchlights had one major problem: "[A]lthough under most favorable circumstances the lights will detect objects at distances up to 2500 meters, they will betray the presence of ship or fleet to the enemy at distances far beyond this range."[15] The article confidently stated that the searchlight would be an indispensable adjunct of flotilla craft, and declared that although "[t]he employment of the electric search light during night engagements between vessels has not yet occurred, . . . it is safe to predict that it will occur."[16] As in the case of torpedoes, navies placed great faith in this emerging technology, acquiring and using searchlights of ever-greater power. They developed elaborate and optimistic tactics for their use, based mostly on theory rather than combat experience. Searchlights also became a reliable means of communicating at night and eased the challenges of nocturnal navigation. These demonstrated benefits, combined with optimistic assumptions about their combat potential, led navies to arm their battleships with large batteries of searchlights. The U.S. Navy's *Arkansas* class of 1912, for example, carried sixteen powerful lights.

Radio

The 1896 invention of radio gave navies the potential to improve their communications further. Unlike searchlights, radio could transmit information over great distances without visually revealing one's position. The promise of improved communications was so alluring that every moderately well-funded navy began buying radio sets, helping to establish a global electronics industry within years of radio's invention.

Early radio sets were delicate, their ranges uncertain, and their transmissions subject to interference—natural and human-generated. Procedures and practice were largely experimental. Some navies appreciated the vulnerability of radio communications—signals could be picked up by any properly tuned receiver within range or jammed by nearby transmitters—but not all. Efforts to conceal important details by using codes and encryption techniques were rudimentary if they existed at all. However, the development of radio was greatly accelerated when a major war erupted during the technology's infancy. The Russo-Japanese War incentivized continued improvement and provided immediate feedback. As a result, in contrast to torpedoes and searchlights, which evolved over long periods of relative peace, radio developed rapidly due to its early use in war.

In fact, naval combat during the Russo-Japanese War became a laboratory for developing all these technologies. European and American attachés and observers closely followed events, giving those nations that did not participate a chance to clarify their expectations for night battle, devise new tactics, and refine their equipment, platforms, and weapons. Advancements in these areas ultimately changed night combat from an uncertain gamble to an advantageous tactic.

The most important factor in this transition was not technological, however; people—their skill, training, and preparation—remained the most crucial determinant of victory or defeat in night action. This axiom was succinctly expressed by Rear Adm. Adolphus Andrews in June 1941: "In the navy we have ships, guns and men;" he wrote, "but the greatest of these is men."[17] Well-prepared officers and trained crews, using their knowledge and skill, made the difference in the dark. As the following chapters will demonstrate, success in night combat required harnessing new technologies and using them to enhance the skill and aggressiveness of talented officers and sailors. Many navies sought victory in night battle, but only those that deliberately invested in enhancing the talents and capabilities of their people consistently achieved it.

POLICY, POLITICS, AND A NAVY'S ROLE

The technological changes that swept through the world's navies in the late nineteenth century occurred within a broader social and political context. It was a time of colonial expansion, as the major powers spread their political and economic hegemony over much of the world. Europe was the center of global military power, and within Europe Great Britain maintained the most powerful navy and possessed the largest colonial empire. By 1904 the British and French empires together comprised 35 percent of the world's land mass. Russia expanded across Asia to the Pacific. Spain, Portugal, Belgium, the Netherlands, and Denmark also possessed extensive colonial holdings. The United States, whose leaders had long professed disinterest in colonial expansion, ousted Spain from its imperial possessions in the Caribbean and the Pacific, thereby becoming a colonial power as well. Germany and Italy, unified late in the colonial era, rushed to acquire their own overseas dominions.

Transoceanic trade was vital to the success of these colonial enterprises, and commerce protection—as well as destruction—was a core mission of the world's larger navies. In the battleship era, it was generally believed that the best way to secure one's own commerce and disrupt that of potential enemies was to command the sea with a battle fleet. Accordingly, Great Britain, Germany, France, Russia, Japan, and the United States all invested in large battle fleets that reflected their commitment to their trading interests and relative standing in the world. Lesser powers, such as Italy, Austria-Hungary, Brazil, Argentina, and Spain, constructed modern dreadnoughts in quest of regional dominance.

Great Britain possessed the largest battle fleet. In the late nineteenth century, its fleet outnumbered the French thirty-four battleships to thirteen. Determined to keep it that way, the British adopted a "two-power standard" in the late 1880s; for decades, their industrial and economic strength delivered a modern battleship fleet "at least equal to the naval strength of any two other countries [combined]."[18].

Britain's rivals chaffed under this asymmetry and sought weapons and tactics to diminish the Royal Navy's superiority. As soon as the torpedo appeared, it was identified as such a weapon. The close blockade, which had been used to keep enemy ships under observation and destroy them if they sortied, appeared obsolete in the face of modern torpedo craft. Torpedo boats could sail from their harbors, fall upon blockading ships under cover of darkness, and sink them. Once a blockading squadron had withdrawn or been driven off, larger ships could sortie to seek battle with crippled enemy ships or to raid enemy commerce.

These tactics, which appeared to "nullify" command of the sea by rendering close blockade impossible, became the centerpiece of a new strategic concept that emerged in France in the 1880s, the Jeune École.[19] Its proponents argued that it was futile to attack British naval superiority by building more battleships, since British economic and industrial superiority would enable it to outbuild any rival. Instead, the key to strategic success was an alternative strategy based on a fleet of fast cruisers and swarms of torpedo boats. The torpedo boats had a dual role. They would harass blockading British battleships and drive them off, freeing the cruisers to attack British merchant shipping. Then, in concert with the cruisers, the torpedo boats would conduct a "sudden paralyzing attack" on seaborne trade to shock the British commercial system, trigger "an economic panic," and "bring about social collapse."[20] Advocates of the Jeune École strategy knew that the British economic system relied on foreign trade—in 1895, for example, almost 80 percent of Britain's wheat was imported—and therefore was vulnerable to maritime disruption.

Like the early proponents of strategic bombing, adherents to the Jeune École overestimated the potential of their preferred technologies and underestimated the capacity of industrial societies to overcome economic disruptions.[21] However, certain aspects of the Jeune École concept reflected the changing nature of naval warfare. Narrow waters and close blockades were becoming increasingly untenable for battleships because of the threat from torpedo craft. Commerce destruction was a viable strategy, even if such attacks were unlikely to trigger a sudden economic collapse. Britain would face these realities a few decades later during World War I. By that time, its Royal Navy had adjusted to the threat offered by the Jeune École and introduced new tactics and ship types, such as the destroyer, the armored cruiser, and the battle cruiser.

American naval historian and strategist Alfred Thayer Mahan regarded the theories of the Jeune École as a threat—not to the United States or its commercial interests, but to the intellectual development of U.S. Navy officers.[22] For Mahan, new technologies would always be secondary to the fundamental principles of naval warfare and the skilled officers who applied them. Technologies forced officers to adapt, adjust, and remain mentally acute, but Mahan believed that because technology was evolving so rapidly in the late nineteenth century, those who overemphasized specific technologies, as the proponents of the Jeune École did, would soon be left behind.

Foremost among Mahan's principles was command of the sea—"the great object of naval warfare"—which he argued was the key to national security and economic prosperity.[23] Mahan emphasized the importance of battle, and stressed its value "to break up the enemy's power on the sea, cutting off his communications . . . , drying up the sources of his wealth in his commerce, and making possible a closure of his ports."[24] Mahan's emphasis on the importance of battle led some adherents to invert the intended relationship; instead of seeing battle as the means to achieving command of the sea, they saw battle as an end in itself.[25]

British historian Julian S. Corbett recognized this weakness and stressed to his readers that, "When we say that the primary object of our battle fleets must always be the destruction of the battle fleets of the enemy, what we really mean is that the primary function of our battle fleets is to seize and prevent the enemy from seizing the main lines of communication."[26]

The difference between Mahan's perspective and Corbett's arose from the differences in their primary audiences. The United States was a rising power, seeking to emerge on the world stage. Great Britain was an established one, with a preeminent navy that others hesitated to contest directly. The Jeune École offered an alternative perspective, one arising from a uniquely French context. These differences illustrate an important truth: each of the navies described in the following chapters developed tactics, doctrine, and techniques for night combat that reflected their unique perspective and strategic circumstances. As the potential threat of torpedo craft increased and navies moved to acquire them, those perspectives shaped how the new ships and their crews were trained, employed, and integrated into existing naval forces. Preparation for delivering night attacks and plans to counter them shared numerous similarities across navies, but each navy solved the associated problems in its own way. The need for solutions became urgent during the Russo-Japanese War of 1904–1905, when modern torpedo craft demonstrated their combat potential on the world stage. That is where our story begins.

CHAPTER 1

STUMBLING IN THE DARK
THE RUSSO-JAPANESE WAR, 1904–1905

STEPHEN MCLAUGHLIN

The night of 8–9 February 1904 was a busy and confusing one for Vice Admiral Oskar Karlovich Stark, commander of Russia's Port Arthur Squadron.[1] Concerned about the possibility of a surprise Japanese attack— most of his ships were anchored in the roadstead outside the harbor—he had convened a conference of senior officers aboard his flagship, the battleship *Petropavlovsk*, to discuss defensive measures. The conference had ended at 2300, and just over half an hour later the bitterly cold winter night suddenly echoed with the sound of an explosion. Within moments gunfire erupted from several of the Russian warships. But *Petropavlovsk* was anchored closer to shore than most of the other ships, and Stark could not see what had happened to seaward; earlier that day he had ordered his destroyers to fit warheads to their torpedoes, and he assumed that someone had gotten careless.

Stark's misperception was not easily dispelled. Apparently believing that his ships were shooting at phantoms in a panic, at midnight he ordered them to cease fire. But the ships kept shooting; in desperation, Stark ordered his searchlights turned skyward, a standard signal for ending a nighttime training exercise. But the gunfire continued.

It was only an hour after the first explosion that Stark finally grasped what had happened: Japanese destroyers had approached under cover of darkness and torpedoed his two best battleships and one of his cruisers. All three wounded vessels would need lengthy repairs, and in a few minutes his squadron had gone from approximate equality with the Japanese battle fleet to a decided inferiority.

TABLE 1.1. COMPARISON OF THE FLEETS

TYPE	JAPAN COMBINED FLEET	RUSSIA PORT ARTHUR SQUADRON	RUSSIA VLADIVOSTOK SQUADRON
Battleships	6	7	—
Armored Cruisers	7	1	3
Protected Cruisers	7	6	1
Destroyers	19	25	—
Torpedo Boats	90	—	11

Notes: Only vessels less than ten years old are included. The Port Arthur Squadron lost two protected cruisers at the outset of the war, while the Japanese received two Italian-built armored cruisers soon after the war began.

Sources: Sergei Suliga, *Korabli Russko-Iaponskoi voiny 1904–1905 gg.* [Ships of the Russo-Japanese War 1904–1905] (Moscow: Askol'd, 1993), 4–35; David C. Evans and Mark R. Peattie, *Kaigun: Strategy, Tactics and Technology in the Imperial Japanese Navy 1887–1941* (Annapolis, MD: Naval Institute Press, 1997), 90–91; *Conway's All the World's Fighting Ships 1860–1905* (New York: Mayflower Books, 1979), 218–39.

PHOTO 1.1. The Japanese destroyer attack on the night of 8–9 February 1904, from an illustration by the British maritime artist Charles Edward Dixon. This drawing, published within days of the event, captures the popular conception of a torpedo attack—the destroyer caught in a searchlight beam as she speeds away from her victim, the torpedo exploding against the hull of a battleship even as she opens fire, while another ship in the background fires a rocket as a distress signal. In reality, the Japanese destroyers launched their torpedoes from hundreds of yards away, and from the port side of the Russian ships, not the starboard. *Author's collection*

THE OPPOSING NAVIES

At the start of the Russo-Japanese War (8 February 1904–5 September 1905) the two navies were fairly evenly matched, as Table 1.1 shows.[2] However, the two Russian squadrons were too far apart to provide mutual support, and their roles in the war would be quite different. The Vladivostok Cruiser Squadron carried out several raids on Japanese communications, but aside from a brief and inconclusive encounter with Japanese torpedo boats on 1 July 1904 it fought no night actions.[3] The Port Arthur Squadron, on the other hand, participated in numerous engagements in the dark.

In terms of individual ships, the Japanese navy's battleships were larger, faster, and better-protected than their Russian counterparts, and its armored cruisers were more modern and more powerful. As for destroyers, Table 1.2 gives the basic characteristics of representative examples of these indispensable craft. Originally conceived as "torpedo boat destroyers" in the early 1890s, the performance of these vessels in the Russo-Japanese War demonstrated that they were jacks-of-all-trades—acting as scouts, escorts, dispatch vessels, blockaders, and as the premier night fighters. Service in these craft was physically demanding. Small and low in the water, the boats were lively in any sort of a seaway, which could be exhausting, while the constant exposure to saltwater spray could inflame eyes.[4]

TABLE 1.2. TYPICAL DESTROYERS OF THE RUSSO-JAPANESE WAR

NATION	NAME	ENTERED SERVICE	DISPL. (NORMAL)	SPEED	GUNS	TORPEDO TUBES
Japan	Shirakumo	1902	342 tons	31 knots	1 × 76.2-mm 5 × 57-mm	2 × 450-mm
Japan	Oboro	1899	305 tons	31 knots	1 × 76.2-mm 5 × 57-mm	2 × 450-mm
Japan	Chidori (1st class torpedo boat)	1900	150 tons	29 knots	1 × 57-mm 2 × 42-mm	3 × 356-mm
Russia	Bezstrashnyi	1900	350 tons	27.4 knots	1 × 75-mm 5 × 47-mm	3 × 381-mm
Russia	Vlastnyi	1902	312 tons	28 knots	1 × 75-mm 5 × 47-mm	2 × 381-mm
Russia	Reshitelnyi	1902	258 tons	27 knots	1 × 75-mm 3 × 47-mm	2 × 381-mm

Note: During the war some Japanese destroyers had a second 76.2-mm gun installed in place of one of the 57-mm guns. (Attaché Reports, 212, report dated 28 September 1904).

Sources: Anthony J. Watts and Brian G. Gordon, The Imperial Japanese Navy (Garden City, NY: Doubleday & Company, Inc., 1971); S. S. Berezhnoi, Kreisera i minonostsy: Spravochnik [Cruisers and Torpedo Craft: A Reference Book] (Moscow: Voennoe izdatel'stvo, 2002); Suliga, Korabli, 30–34.

In night actions, torpedoes were their primary weapon. In both navies, each destroyer carried two or three single torpedo tubes, usually arranged on the centerline. Most destroyers also carried several torpedo reloads, usually one per tube.[5] The Japanese also had large numbers of torpedo boats, smaller and shorter-ranged than destroyers.

Since the ships of both navies were equipped with the latest developments, the war would serve as the proving ground for several new technologies that would shape naval warfare in the coming decades. The most important of these for night fighting were:

Rapid-Fire Guns: These were used on flotilla craft, cruisers, and capital ships. On the lighter vessels they were the main armament; on the larger ships they were the first line of defense against the fast and stealthy torpedo boats. The most common calibers were 37-mm (also called 1-pounders after the weight of the shell they fired), 47-mm (3-pounders), and 57-mm (6-pounders); only the Japanese used the 57-mm size. The Russians also had 75-mm rapid-fire guns, while the Japanese had 76.2-mm (3-inch) weapons. Neither fleet possessed special optical gear for use at night, but illuminated front sights for guns were available, so a gunner could at least see where his muzzle was pointed.[6]

Torpedoes: Both navies used torpedoes either purchased from the Whitehead company of Fiume or built domestically under license (see Table 1.3). Most, if not all, torpedoes in both navies were fitted with the Obry gyroscope, which greatly improved their accuracy.[7] Russian warheads used pyroxilin (wet guncotton) as their explosive, while the Japanese used the more powerful Shimose, a form of picric acid.[8]

Torpedoes were launched from their tubes by a gunpowder charge, making a flash that could give warning to an intended victim; for example, during the night after Tsushima the surviving Russian cruisers of Rear Admiral

TABLE 1.3. TORPEDOES OF THE RUSSO-JAPANESE WAR

NATION	DESIGNATION	CALIBER	WEIGHT	WARHEAD	MAX. RANGE @ SPEED
Japan	Type 32	450 mm	541 kg	90 kg	1,000 m @ 28 knots 3,000 m @ 15 knots
Japan	Type 32	356 mm	338 kg	50 kg	800 m @ 24 knots 2,500 m @ 15 knots
Russia	Model 1898 L	381 mm	430 kg	64 kg	600 m @ 30 knots 900 m @ 25 knots

Note: Japanese destroyers carried the 450-mm torpedo, torpedo boats the 356-mm.

Sources: Lacroix and Wells, 778; Korshunov and Uspenskii, 26.

Oskar Adolfovich Enkvist's detachment noted that the Japanese torpedo craft were "sometimes so close that we could clearly hear the sound of torpedoes being launched from their tubes, and see the flashes of their powder charges."[9]

Searchlights: Searchlights had been standard equipment for warships since the 1870s. They used a bright arc light reflected off a large-diameter mirror to cast a beam of intense light. By the 1880s they could pick up an approaching ship at more than two thousand yards, and the crews of vessels caught in their beams usually would be temporarily blinded.[10] Both Russian and Japanese searchlights had Mangin reflectors, which produced a more-focused beam than a parabolic mirror did; at a range of a thousand meters it had a width of about thirty meters.[11]

The standard Russian searchlight had a diameter of 75 cm. Battleships and large cruisers generally had six searchlights, providing for all-around coverage; smaller cruisers were equipped with three or four; and destroyers usually had a single light.[12]

Japanese battleships had five or six 60-cm searchlights; details on the outfit of other types of ship are lacking, but probably were similar to those of their Russian counterparts.[13]

PHOTO 1.2. The Japanese destroyer *Kasumi*. Built in Britain by Yarrow, a shipyard famous for its fast torpedo boats and destroyers, she made more than thirty knots on trials. In this photo, she still lacks her armament—one 3-inch gun on the "bandstand" forward of the conning tower, five 57-mm guns and two 18-inch torpedo tubes. Delivered in June 1902, she saw much action during the war with Russia, and was credited with the torpedo hit on the cruiser *Pallada* during the initial attack at Port Arthur. She was stricken from the active list in 1913 and served as a target ship until sometime in the 1920s. *Bain News Service, ca. 1910–15; George Grantham Bain Colletion, Library of Congress*

Radio: Both navies had installed radios in all warships larger than destroy-ers, while the Japanese would begin equipping destroyers with radio sets starting in the summer of 1904.[14] But the existing "spark" radios could send only Morse-like signals and therefore were slow; they also were easily jammed. Range was uncertain. When the Russian Port Arthur Squadron went out on a short cruise on 3–4 February 1904, the minelayers *Amur* and *Yenisei* trailed fifteen and thirty miles respectively behind the squadron, forming a commu-nications chain with the main base.[15]

Although the characteristics of ships and weapons are readily available in reference books, far less is known about the tactical doctrine of the two forces—that is, how the Japanese and Russian navies expected to use the new weapons. In Japan, "the ultimate formulation of officially sanctioned tactical thinking" was the highly secret *Kaisen yōmurei* (Battle Instructions).[16] Unfor-tunately, what the instructions may have said about night actions is unknown, but Vice Admiral Tōgō Heihachirō, the commander in chief of the Combined Fleet, did define the role of the destroyers and torpedo boats as a part of his fleet: they were to stay out of the way during a daylight gunnery action, attacking only if they saw a favorable opportunity, and they were to pursue the defeated enemy at night, "taking advantage of darkness to blow them up."[17] Lieutenant Commander Akiyama Saneyuki, who joined Tōgō's staff in 1903, was a particularly important innovator in both strategy and tactics and played a role in formulating the fleet's doctrine.[18] As such, he may also have contributed to a change in torpedo doctrine just before the outbreak of the war, from close-range attacks to long-range ones. Night attacks at short ranges had been used to disable the Chinese battleship *Ting Yuan* and the cruiser *Lai Yuan* at Weihaiwei during the Sino-Japanese War of July 1894–April 1895; it was a method that accorded well with the ancient Samurai concept of *nikuhaku-hitchū*—"press closely, strike home."[19] But starting in 1902 the Japanese navy began a series of torpedo exercises that eventually led to an emphasis on firing torpedoes at long ranges using the low-speed setting—ironically, a concept borrowed from the writings of Russian Admiral Stepan Osipovich Makarov.[20] This policy was in place during 1904—a fact confirmed by British officers attached to the Japanese fleet, who noted that torpedoes were usually set to run three thousand meters, limiting their speed to fifteen knots.[21] Japanese doctrine apparently also prescribed that torpedo craft should attack individually rather than in groups, reportedly to avoid colliding with one another.[22] Julian Corbett, author of the most penetrating study of the war, noted "Even when a [destroyer or torpedo boat] division came across a group of the enemy, it would break up, each boat attacking a different ship. Not once was there an attempt to push home in a bewildering mass."[23] Judging by wartime events, it was also standard practice to fire a single torpedo at each target.

The state of the Russian navy in terms of doctrine and practice is less clear. There was no navywide document equivalent to the *Kaisen yōmurei*; individual commanding admirals were responsible for making their intentions for battle clear to their subordinates under Article 107 of the Naval Regulations.[24] Russian experience in the use of torpedo boats dated to the Russo-Turkish War of 1877–78, when launches armed with spar torpedoes were used in night attacks against Turkish warships on the Danube River and the Black Sea.[25] A harbinger of the future came when Makarov, then a young lieutenant, led an operation that saw the first successful use of a Whitehead torpedo, sinking the Turkish gunboat *Intibah* on the night of 26 January 1878.[26] Torpedo boats had gone on to form an important element in Russian naval planning throughout the 1880s and 1890s; by the turn of the century, the general view was that "nighttime attacks with the successive launching of several torpedoes at intervals of not less than ten seconds" would be the best tactic when attacking ships at anchor or those damaged during a daylight gunnery action.[27] As for defensive measures, in 1898 the commander of the Pacific Squadron had established a commission to review the existing instructions for repelling torpedo attacks, including night attacks; unfortunately, little is known of the result, although the use of searchlights to scan for approaching torpedo boats was subjected to extensive study.[28] As in other navies, when night attacks were expected the crews of the rapid-firing guns slept near their guns and extra officers were detailed to maintain the watch. Night recognition signals used red and white colored lights—coincidentally the same colors used by the Japanese.[29] Flashing bull's-eye lanterns were used for night-signaling by destroyers.[30]

In sum, the two navies had similar ships and equipment and both had some form of doctrine—but the two were by no means equal in terms of training and spirit. This is a topic fraught with intangibles and prejudice, but there is little doubt that Russian naval officers were in general less capable than their Japanese counterparts.[31] This was due in part to a general shortage of officers, a longstanding problem in the rapidly expanding Russian navy; in late 1903 the Port Arthur Squadron lacked ninety-six officers, and the problem was particularly severe among the destroyer crews.[32] This contributed to the high turnover rate among Russian destroyer captains and division leaders: after six months of war, only two boats had the same commanders with whom they had started, and some of these officers had little previous experience with torpedo vessels.[33] In a number of cases, captains served only a few weeks—or even days—before being replaced.[34] As a result, commanders had little time to learn their jobs or to train their crews to be effective teams. Many also appear to have been too old for their assignments; one observer claimed that the average age of destroyer commanders was about 40, and that many could not withstand the strain of serving in these small vessels for long.[35]

Compared with their Russian counterparts, Japanese destroyer and torpedo boat commanders were somewhat younger and had considerably more experience. All had previously served in torpedo craft or as torpedo officers in larger ships, and most (63 percent) had spent considerable time in such positions.[36] As one of the British attachés noted, "In the Japanese Navy . . . service in torpedo vessels is almost a speciality. Commander Mano informed me that he had been over 10 years in command of torpedo boats; he was in command of one of the boats that attacked and torpedoed the Chinese battleship in Weihaiwei in 1895, and has been in no other type of vessel since. This is by no means an isolated case, and accounts in some measure for the admirable handling of these small craft."[37] However, in the months between the battles of the Yellow Sea and Tsushima many destroyer commanders were replaced by younger men promoted from torpedo boats; perhaps the older Japanese officers, like their Russian counterparts, were finding the strain of prolonged wartime service in these small vessels too much.[38]

THE PORT ARTHUR ATTACK, 8–9 FEBRUARY 1904

By early 1904 the incompatible interests of Russia and Japan in Manchuria and Korea had led to a crisis. Both sides expected war to break out in the near future, and both realized that the success of the land campaigns would hinge on naval power. Because there were no suitable landing places on the eastern coast of the Korean peninsula, the Russians expected the Japanese to carry out amphibious landings on the west coast, then march northward into Manchuria, cutting off the naval base at Port Arthur on the Liaodong Peninsula. But the Russians would be in an excellent position to interfere with Japanese maritime communications along Korea's west coast. They therefore planned

MAP 1.1. OVERVIEW OF RUSSO-JAPANESE WAR NAVAL OPERATIONS

to use their squadron to prevent or disrupt Japanese landings, thereby giving their vast empire's scattered military forces time to concentrate and defeat the Japanese army.

As a result, from the Russian perspective the Port Arthur Squadron was a "fast reaction force" that had to be ready to put to sea at a moment's notice to counter any Japanese moves. This imperative explains the posture assumed by the Russian command on the eve of the war. Most critically, starting in January the Russian squadron's normal anchorage was not in Port Arthur's inner harbor, but in the outer roadstead. The channel into the harbor was so shallow that big ships could only pass through it at high tide, so half a day might pass before a squadron in the harbor could get to sea to counter a Japanese landing. Moreover, the channel was narrow and winding, and there were fears that the Japanese might attempt to block it, thereby bottling up the squadron. At the same time, the Russians were well aware of the vulnerability of the outer roadstead to a surprise attack, having studied it during a large-scale wargame played in St. Petersburg over the winter of 1902–1903. It was also noted at a conference in April 1903, although at that time Rear Admiral V. K. Vitgeft, the naval chief of staff to the viceroy of the Far East, downplayed the risk because the ships would be protected by torpedo nets.[39]

But as war approached, the squadron's commander, Vice Admiral Stark, ordered that the nets not be rigged out, and in fact chastised several ships that attempted to put them out without orders.[40] This, too, was the result of the squadron's role as a fast-reaction force; in a report prepared only hours before the attack, Stark said that having to take the nets in would "delay the movement of the squadron when it needs to weigh anchor in an emergency."[41] Instead, there were plans to create a boom defense for the roadstead using nets captured from the Chinese at Taku during the Boxer Rebellion.[42] However, the boom was not yet ready when the Japanese attacked. Moreover, the fortress of Port Arthur, including its coast-defense batteries, was not put on alert, and the Laotieshan Lighthouse (about five nautical miles southwest of Port Arthur) continued in operation.

Not all precautions were neglected. Every night Stark signaled "prepare to repel torpedo attack," and the crews of the rapid-firing guns slept at their guns, which were kept loaded. Two cruisers were to have steam up so that they could react immediately if the need arose. Two destroyers were sent twenty miles out to sea every night to patrol, but because they were not equipped with radios they had to return to the flagship to report if they saw anything suspicious. Two ships were detailed for searchlight duty every night, while the other ships were to be darkened, although bow and stern lights were to be shown.[43]

But these measures were compromised by orders to keep the ships fully coaled—another consequence of the quick-reaction policy. So every night several ships had barges alongside and lamps rigged on deck as they topped up their bunkers. Moreover, some officers assumed that the precautions were an exercise, in part because the torpedo nets were not used, and so some of the ships did the minimum to comply with Stark's orders.[44] Security was so lax that an American reporter, Frederick McCormick, was able to hire a boat and cruise among the squadron's ships "without being seriously challenged" only a few hours before the Japanese attack.[45]

Thus, on the night of 8–9 February 1904 the major units of the Russian squadron were anchored without torpedo nets in the outer roadstead in three east-west lines, with their bows pointing southwest (see Map 1.1). A bit farther out was the auxiliary cruiser *Angara*, while the small cruiser *Novik* and the gunboats *Giliak* and *Dzhigit* were anchored in another line inshore of the big ships.

Japanese plans mirrored Russian expectations almost exactly. Admiral Tōgō had argued for both an attempt to block the channel to the inner harbor and for a surprise torpedo attack on any Russian ships outside Port Arthur.[46] The purpose of an attack before the declaration of war was the same as that of Pearl Harbor thirty-seven years later: to paralyze the enemy's navy so that it could not interfere as Japanese forces seized their objectives. The Naval General Staff approved the idea, and when the fleet sailed from Sasebo on 6 February 1904 it was led by the five divisions of destroyers that were earmarked for the attack. While en route Tōgō received an intelligence report indicating that while the bulk of the Russian squadron was anchored in Port Arthur's roadstead, three battleships had put to sea for parts unknown. This information probably came from the Japanese consulate at Chefoo (today the Zhifu District of Yantai), "the prime source of information about the movements of naval ships in and out of Port Arthur."[47] By about 1700 on 8 February the Japanese fleet was approaching Round Island (today Yuán Dǎo), about forty-five miles east of Port Arthur, and Tōgō detached the destroyers to carry out their mission. Three divisions were assigned to attack the ships in Port Arthur's roadstead, while two divisions were sent to Dalny (now Dalian), an undefended commercial port about twenty miles up the coast.

The destroyers involved were:[48]

Port Arthur Attack

- ▸ First Division: *Shirakumo, Asashio, Kasumi, Akatsuki*
- ▸ Second Division: *Ikazuchi, Oboro, Inazuma* (a fourth boat, *Akebono*, had been damaged in a collision the day before and did not take part in the attack)
- ▸ Third Division: *Usugumo, Shinonome, Sazanami*

Dalny Attack

- ▸ Fourth Division: *Hayadori, Harusame, Murasame, Asagiri*
- ▸ Fifth Division: *Kagerō, Murakumo, Yūgiri, Shiranui.*

Tōgō apparently reasoned that the three battleships that had reportedly sailed from Port Arthur were at Dalny, and so he assigned nearly half his strike force to it, with twenty torpedo tubes in the Port Arthur group and sixteen tubes in the Dalny group.[49] But in fact there were no Russian ships at Dalny, so the eight destroyers ordered there accomplished nothing—although they very nearly torpedoed a British steamer evacuating Japanese residents from that town. Tōgō had seriously weakened the strike force bound for Port Arthur, where he knew the main body of the Russian squadron was located, on the basis of a vague intelligence report.

The Port Arthur group meanwhile proceeded in a single column with the First Division leading.[50] Although it was bitterly cold, the sea was calm, the weather clear, and the moon would not rise until 0035, so until then they could count on profound darkness. It was, therefore, "an almost perfect night for the operation."[51] At about 2135, while still some distance from their target, they sighted lights ahead that they took to be Russian searchlights, helping them to home in on their targets. Captain Asai Sojiro of the First Division, in overall command, had planned a coordinated attack, with his division turning to port to make its run while the other two divisions turned to starboard. But about twenty minutes after sighting the searchlights the Japanese formation was thrown into confusion when two new lights were sighted approaching on their port bow. These were the Russian sentry destroyers *Bezstrashnyi* and *Rastoropnyi*, showing their usual peacetime lights; they were on a course to cut right through the Japanese line. The First Division altered course slightly to starboard and extinguished its stern lights; the *Ikazuchi*, leading the Second Division, turned to starboard and slowed down to avoid the Russians, but the following destroyer, *Oboro*, failed to notice her leader's actions and held on, colliding with her. *Ikazuchi* sustained only minor damage and continued on her way, but *Oboro*'s bow was more seriously damaged, forcing her to stop, while the third boat, *Inazuma*, also stopped to render assistance. The Third Division, bringing up the rear, came to a complete halt and lost touch with the other two divisions, while the last boat of the division, *Sazanami*, mistook the Russian destroyers' stern lights for those of her comrades and fell in astern of them. Asai's plan for a coordinated attack was ruined.

Meanwhile the two Russian destroyers, having passed through the Japanese formation about two hundred yards astern of the trailing boat of the First Division, continued on their way, unaware of the chaos their appearance had caused; they apparently never saw the Japanese destroyers, and would return to the squadron to make their scheduled report in the midst of the Japanese attacks.[52]

By 2325 the four boats of the First Division were approaching the Russian anchorage and Asai ordered them to attack. He estimated the range as six hundred meters, while the other boats judged it to be four hundred to five hundred meters.[53] Turning to port in succession, they traversed the Russian squadron's front, launching a total of seven torpedoes.[54] Near the end of this attack the

PHOTO 1.3. The Russian destroyer *Burnyi*. Built by the Nevskii Works to a variation of a Yarrow design—note the similarity of her profile to that of *Kasumi*—she entered service in 1902. Unlike *Kasumi*, however, she had a fixed 15-inch torpedo tube in her bow as well as the two trainable tubes aft of her funnels. The 75-mm gun on the bandstand forward is plainly visible, but her five 47-mm guns are lost in the topside clutter. She was active in the defense of Port Arthur. After the Battle of the Yellow Sea, she ran aground on the Shantung Peninsula in foggy weather and was destroyed by her crew. *V. Ya. Krestyaninov, SPb., 2006 via Wikimedia Commons*

Russians opened fire and some of the boats were hit, but none was seriously damaged and there were no casualties. The first attack had lasted about seven minutes (2333–2340), and several explosions were noted among the Russian ships. The First Division then withdrew to the south at high speed.

Ikazuchi, having approached independently, made her attack hard on the heels of Asai's division. Having turned to starboard as previously arranged, at 2340 she fired both her torpedoes at an estimated range of a thousand meters. She came under heavy but inaccurate fire and was able to retire southward. She did not see any results from her attack.

The Third Division, after becoming separated from the others, soon came upon the damaged *Oboro* and her division-mate, *Inazuma*. The latter requested and received permission to join the division, and the three ships proceeded toward the lights of the Russian squadron, *Inazuma* in effect replacing *Sazanami*, which had disappeared. They approached to within about fifteen hundred meters of the Russian squadron then launched all six of their torpedoes. Their attack lasted about six minutes (2344–2350).

There were two more attacks, made by single boats. *Sazanami* eventually realized the boats she was following were not Japanese, so she broke away and decided to proceed on her own. Guided by the Russian searchlights, she approached the anchorage at low speed. When the searchlights were suddenly directed skyward—Stark's attempt to order cease-fire—she was able to close to an estimated range of seven hundred meters, and at 0030 she launched both her torpedoes and then made off in the darkness. *Oboro*, meanwhile, having determined that her damage was not as serious as first thought, also decided to make an independent attack; she found the Russian squadron, now darkened, about twenty minutes after *Sazanami* and launched her torpedoes at a range of about twelve hundred meters, then withdrew. After their attacks all the destroyers proceeded to their rendezvous point at Round Island.

Captain E. C. T. Troubridge, a British attaché with the Japanese fleet who interviewed several of the Japanese officers involved, noted the "difficulty of reconciling the various accounts of torpedo attacks," and that indeed was the case.[55] The destroyer commanders considered their attack a resounding success; they had fired a total of nineteen torpedoes, and Asai believed that at least five, and perhaps as many as seven, torpedoes had scored hits, while the commander of *Akatsuki* was certain that eight torpedoes had hit.[56] In fact, only three ships had been damaged: the battleship *Retvizan* was hit on the port side forward a little after 2335, the battleship *Tsesarevich* was struck on the port side aft at about 2338, and the cruiser *Pallada* was torpedoed on the port side amidships soon after 2345. *Retvizan* and *Pallada* were on searchlight duty that night, making them prominent aiming-points for the Japanese destroyers.[57]

Thus all the hits came within a span of about ten minutes, and were very probably scored by the First Division, the only one that claimed hits in its after-action report.[58] After their initial confusion, the Russian ships opened a vigorous fire on the attackers, and over the course of the attacks shot off approximately eight hundred shells, ranging in caliber from 37-mm to 6-inch.[59] Although the Japanese destroyers sustained little damage and no casualties as a result of this barrage (*Akatsuki*, the most heavily damaged boat, was hit only twice), it seems to have deterred some of the later arrivals, most of whom reportedly fired from ranges of more than a thousand yards.[60] The Japanese torpedoes were also fitted with net-cutters, which would have reduced both speed and accuracy.[61]

The human factor also must be taken into account: many of the men were going into battle for the first time, and several Japanese officers admitted afterward to feeling "off balance" and to "firing blind."[62] For men who were excited or nervous it would have been all too easy to make mistakes: ranges were almost certainly underestimated, and in one of the ten unexploded torpedoes recovered by the Russians after the action the safety pin had not been removed from the detonator.[63]

Despite careful planning, extensive training, and an unprepared enemy, the attack had been a confused muddle that had achieved considerably less than had been hoped. Nor was this a fluke; future actions would soon demonstrate that night combat was far more difficult than prewar planners had anticipated.

NIGHT DESTROYER ACTION, 10 MARCH 1904

Despite its disappointing results, the torpedo attack had achieved its purpose: with two of its battleships *hors de combat*, the Russian command considered a direct challenge to the Japanese fleet impossible. Skirmishing continued, leading to some minor Russian losses. But the arrival of Vice Admiral Makarov, regarded as Russia's most active and able naval commander, on 8 March buoyed the spirits of the squadron. Nor was Makarov's reputation exaggerated; the day after his arrival he took steps to counter Japanese activities outside Port Arthur. Hoping to catch the enemy off guard, on the evening of 9 March he sent the destroyers *Reshitelnyi* and *Steregushchii* out to discover where the enemy destroyers were based—a mission that would end in tragedy the following morning when, after several brief nighttime encounters with Japanese vessels, *Steregushchii* was sunk in an unequal battle. *Reshitelnyi* was able to escape to Port Arthur.

Although *Steregushchii*'s valiant last stand became a legend in the Russian navy, there was a second, more interesting engagement that night.[64] That same evening the Japanese First Destroyer Division (*Shirakumo, Asashio, Kasumi,* and *Akatsuki*) approached Port Arthur, hunting for any Russian vessels that it might encounter. Finding none in the outer roadstead, it moved down the coast and established a patrol line off the southern tip of the Liaodong Peninsula, with the intention of cruising back and forth on an east-west line about four miles long, hoping to catch any Russian vessels that happened to be at sea.

At about 0100 Russian observation posts ashore had reported seeing lights in the roadstead; Makarov, assuming these were enemy destroyers, decided to take action. He summoned Captain First Rank N. A. Matusevich, commander of the First Destroyer Division, and ordered him to take his four boats out with straightforward instructions: in the event of encountering the enemy, attack him.[65] The Russian force consisted of these ships:

- *Vynoslivyi* (flying Matusevich's broad pennant): Lieutenant P. A. Rikhter;
- *Vlastnyi*: Lieutenant V. A. Kartsov;
- *Vnimatelnyi*: Lieutenant I. V. Stetsenko;
- *Bezstrashnyi*: Lieutenant I. I. Skorokhodov.

Matusevich led his group southwestward, hugging the coast. The moon rose at 0200, but against the dark shore his boats remained practically invisible. As they approached the tip of the peninsula, the Russians noted flashing lights out to sea. These came from the Japanese destroyer *Kasumi*, which was somewhat carelessly signaling to her comrades. The four Japanese destroyers were on the eastward leg of their beat, and Asai—the same officer who had commanded the division during the attack on the Port Arthur Squadron at the war's outbreak—fearing that *Kasumi*'s signals could be seen by any nearby enemy, decided to move closer to shore, ordering the ships to turn back to the west in succession. *Shirakumo* and *Asashio* had already completed their turn, *Kasumi* was still in the process of turning, and *Akatsuki* was still on the eastward track when all hell suddenly broke loose.

Matusevich had made two simple signals to his group: *I see the enemy to port* and *Attack*.[66] He immediately turned toward the Japanese and dashed in at full speed, the other ships following closely. The Japanese had not seen the Russian destroyers against the background of the coast, and their sudden onslaught caught them by surprise.

With the Russians attacking from astern, *Shirakumo* found her fire masked by *Asashio*; Asai ordered full speed ahead in order to get to a more favorable position, but he never found one, and *Shirakumo* took no further part in the action. So while *Vynoslivyi* thought she was engaging the enemy's leader, it must actually have been the second ship, *Asashio*. Cutting across her stern, *Vynoslivyi* fought her at short range—"almost close enough to throw a grenade"—on her port side.[67] *Asashio* sustained eight hits, including one that lodged under a boiler, but none of the Russian shells exploded, which seems to have been the case with most of their shells on this occasion. The *Asashio's* shells were more effective, however, and *Vynoslivyi* staggered back across her bow, her port engine damaged and steam pipes burst. Immobilized, she fired at the other Japanese ships as they passed astern and received their fire in return. At 0345 a shell struck her conning tower, wounding Captain Matusevich. Thanks to exceptional efforts on the part of her engineering crew—Engineer-Mechanic Blinov's hands were severely scalded, but he continued to direct the repairs—at 0355 *Vynoslivyi* was able to steam off at 5 knots. She was out of the action—as was *Asashio*, which had turned to port to avoid a collision when *Vynoslivyi* crossed ahead of her and soon lost sight of the battle.

Lieutenant Kartsov in *Vlastnyi* had decided on a more direct form of attack—ramming. He caught *Kasumi* while she was coming out of her turn and headed straight for her. *Kasumi* was illuminated by moonlight, and she could not at first see her assailant; she stopped to search for the enemy, then accelerated again, finally seeing the Russian ship when she was only two to three hundred meters away. These speed fluctuations were enough to frustrate Kartsov's plan; assuming that *Kasumi* meant to let him cross her bows so that she could ram him, he altered course to cut astern of the Japanese destroyer at a distance of only "five or six fathoms [ten or twelve yards]."[68]

The two ships exchanged fire at close ranges until *Kasumi* saw *Akatsuki*, the last boat in the Japanese formation, send up a rocket as a distress signal.

Kasumi tried to turn about to assist her comrade, but was unable to break away from her opponent at first. Here the reports of the two sides become confused—and confusing. According to the Russian version, *Vlastnyi* fired two torpedoes at the enemy, one of which was seen to explode abreast of the Japanese ship's aftermost funnel. Steam erupted from the target; she heeled over to starboard and began to settle by the stern, fired a rocket as a distress signal, and then sank. But it is clear that it was *Akatsuki*, the fourth Japanese destroyer—not *Kasumi*—that fired a rocket. Somehow in the confusion of action it appears that *Vlastnyi* had fallen behind *Kasumi* and began engaging *Akatsuki*. From *Kasumi*'s viewpoint, the Russian had withdrawn because she had suffered damage, and so *Kasumi*, which by now had lost sight of her comrades, retired from the scene as well.

According to *Akatsuki*, she first engaged a single destroyer, probably *Vlastnyi*, which fired a torpedo at her starboard side; the enemy then cut across her stern, only three or four meters away—suspiciously close to the five or six fathoms by which *Vlastnyi* reported she missed her first opponent's stern. The two ships then fought side-by-side until *Akatsuki*'s opponent drew off due to damage. *Akatsuki* then came under fire from two Russian destroyers on her port quarter—probably *Vnimatelnyi* and *Bezstrashnyi*, the two trailing Russian ships. One of them scored a hit, disabling *Akatsuki*'s engines, and she lost power, so she fired the rocket distress signal that was seen by *Kasumi* and *Vlastnyi*. *Akatsuki* then reports that she was surrounded by enemy ships, but these soon began firing at one another and *Akatsuki* was able to limp away.

Vlastnyi had indeed been damaged during the action, and her steering gear was knocked out, but she was able to maneuver by rigging steering tackle. She reported fighting off more enemies, who eventually turned off their lights and disappeared.

Aside from harassing *Akatsuki*, both *Vnimatelnyi* and *Bezstrashnyi* seem to have played only a supporting role; the official Russian history credits them with no more than driving away the Japanese boats that were threatening the immobile *Vynoslivyi*. It thus appears that the battle had primarily been fought between *Vynoslivyi* and *Vlastnyi* on the one hand, and *Asashio*, *Kasumi*, and *Akatsuki* on the other. This fact is reflected in the casualties sustained: *Vynoslivyi* and *Vlastnyi* between them reported three men killed and twenty wounded, while *Asashio*, *Kasumi*, and *Akatsuki* counted six killed and eight wounded.[69] The other two Russian destroyers and *Shirakumo* suffered no losses at all. The fact that the Russian 75-mm shells did not detonate undoubt-edly reduced the damage and casualties sustained by the Japanese ships. This was a problem that plagued Russian shells of all calibers during the war, but in the case of the destroyers' 75-mm guns there was a known cause.[70] For reasons that defy explanation, only armor-piercing shells, rather than more-effective high-explosive rounds, were available for these guns, and their fuzes were too insensitive to be triggered by the thin plating of destroyer hulls. High-explo-sive 75-mm shells only entered service in 1905.[71]

The Russians did suffer one final mishap. When they had regrouped, Matusevich ordered *Vnimatelnyi* to take *Vynoslivyi*, the most heavily damaged of the destroyers, in tow. But *Vnimatelnyi* miscalculated the turn and rammed *Vynoslivyi* on the port side amidships, flooding an engine room. The collision, as one writer noted, "was due to poor tactical training—the result of insuffi-cient steaming in peace time."[72] In the end, *Vynoslivyi* proceeded under her own power, and the detachment returned to Port Arthur at 0700. Meanwhile, the Japanese destroyers rejoined their fleet and transferred their wounded to the better-equipped battleships for treatment.

Captain Matusevich initially had reported engaging four enemy boats, but after reviewing the reports from the other destroyers was convinced that there had been "not less than six."[73] The Japanese also believed they had engaged a superior force, and both sides thought they had sunk one of the enemy. In fact, however, the two forces were equal in number, and no ships were lost on either side. Although the British official history stated that the battle ended when the Russians withdrew, and one Japanese author claimed that the battle was a "signal defeat" for the Russians, it was really a draw; neither side sought to renew the action, and both withdrew of their own volition.[74]

Vlastnyi's unrealized change of target from *Kasumi* to *Akatsuki* well illus-trates the confused nature of fast-paced night actions. Mistakes in identifying enemy vessels are easy enough to make in daytime; at night the problem is far more difficult. The darkness also explains why both sides lost cohesion at the outset of the battle; ships engaged their opposite numbers and soon lost track of where other vessels, both friends and enemies, were. Another factor to be borne in mind when trying to untangle the course of the action is that the after-action reports had to be compiled from memory, since destroyers had small crews and could not spare someone to record events as they happened. A brief and intense action such as this one—it had lasted twenty minutes at most—was so crowded with crises and rapid decisions that some things inevitably were forgotten or misremembered by the time reports were being written.[75]

THE PORT ARTHUR SQUADRON, FEBRUARY–DECEMBER 1904

Skirmishing between Russian and Japanese destroyers continued, and it was one such encounter that led to the death of Admiral Makarov. On the night of 12–13 April 1904 a group of Japanese destroyers laid mines outside the port. That same night, Makarov sent out a group of destroyers to reconnoiter the Elliot Islands (today Wàichángshān Qúndǎo) and one of the ships, *Strashnyi*, got separated and fell in with some Japanese destroyers, believing them to be her comrades. When morning came, both the Russians and Japanese discovered what had happened, and the outnumbered *Strashnyi* was soon overwhelmed. This had all occurred within sight of Port Arthur, and the cruiser on duty, *Baian*, went to her assistance. Too late to save the luckless *Strashnyi*, she chased off the Japanese destroyers and rescued survivors, but soon the Japanese armored cruisers *Asama* and *Tokiwa*, with four smaller cruisers, appeared. The Russians in turn sent reinforcements, including battleships *Petropavlovsk*, flying Makarov's flag, and *Poltava*. But now the Japanese battle fleet appeared, and in retreating *Petropavlovsk* ran into the minefield planted by the Japanese destroyers during the night. Striking one or two mines, she exploded and quickly sank, taking Makarov and most of her crew with her.

This was a severe blow to the Russians; the energetic Makarov was replaced by the lackluster Rear Admiral Vitgeft, former chief of staff to the viceroy of the Far East. With Port Arthur now besieged by the Japanese army, on 23 June Vitgeft reluctantly took the squadron to sea in hopes of defeating an inferior Japanese force or escaping to Vladivostok.[76] Admiral Tōgō was reluctant to engage in a slugging match, having lost the battleships *Yashima* and *Hatsuse* to Russian mines on 15 May, while the cautious Vitgeft was worried about the threat posed by the numerous Japanese torpedo craft. Having been delayed in putting to sea, he now feared being caught in open waters at night by the enemy flotillas. So, at 1850, he turned the squadron back toward Port Arthur, despite the fact that it would have to remain in the outer roadstead throughout the night because the high tide that his large ships needed to enter the harbor was several hours away.

The Japanese attacks began an hour before the squadron had reached the anchorage, at about 2043.[77] The Russian cruiser *Baian* was leading, followed by the battleships, while the cruisers *Pallada* and *Novik* brought up the rear and bore the brunt of the initial attacks. The only casualty from these was the Japanese torpedo boat *Chidori*, which had been damaged by a friendly torpedo; none of the Russian ships had launched any. Several more divisions of torpedo boats and destroyers attacked before the squadron reached the roadstead, but no hits were scored on the Russian ships.

Vitgeft had signaled to the port commander to switch off the controlled mines as he approached the anchorage, but after months of minelaying by both sides the entire area was strewn with drifting mines, and at 2135 *Sevastopol* was rocked by an underwater explosion on her unengaged side. Her damage was not fatal, and she anchored in the shallow waters of White Wolf Bay, an indentation in the coast a few miles southwest of Port Arthur.

Baian anchored at 2145, the ships in her wake following suit as they came into position, forming a long single line that left some of the ships outside the roadstead's boom defenses. As they anchored, they rigged out torpedo nets. During the subsequent attacks the ships avoided using their searchlights, instead relying on those on the hills on the east side of the harbor entrance; as opportunities arose, the shore batteries also joined in the action. The ships anchored with their bows facing eastward—the direction of the Japanese attacks—thus reducing the size of the targets they offered, although it also reduced the number of rapid-firing guns that they could bring to bear on the enemy.

The attacks continued until 0330 on 24 June. During the entire course of the night ten destroyers and twenty torpedo boats launched a total of thirty-eight torpedoes, most of them at stationary ships, and not a single hit was scored—not counting the one on the unfortunate *Chidori*. The attacks were disorganized, several carried out by individual boats that had lost sight of their comrades in the darkness. On the other hand, the Russians fired 3,072 shells ranging in caliber from 12-inch to 37-mm, scoring a handful of hits that caused minimal damage. However, the heavy fire probably acted as a deterrent; according to Russian reports, most of the Japanese ships attacked from very long ranges—as much as twenty-four hundred yards—and after the action they found twelve torpedoes floating in the roadstead or on the nearby shore.

Once again both sides overestimated the damage they had inflicted. Although no Japanese vessels had been seriously damaged, several Russian ships reported sinking a torpedo boat, in some cases providing convincing details. Vitgeft, although skeptical of some of these reports, nevertheless thought that three boats had been sunk. This belief was reinforced when the Russians discovered the bodies of a Japanese officer and three seamen washed ashore in the morning. For their part, the Japanese thought that two battleships, *Sevastopol* and either *Peresvet* or her sister *Pobeda*, had to be beached, and that a *Diana*-class cruiser had to be towed into the harbor.

Over the next weeks Russian losses continued. One unusual incident came on the night of 23–24 July 1904, when three Russian destroyers were anchored in Ta Ho Bay, a few miles east of Port Arthur, where they had been guarding the army's right flank.[78] The Japanese staged a demonstration to seaward, and, while the Russian ships were occupied with that, torpedo launches—steam-powered boats from the battleships *Mikasa* and *Fuji* that could be equipped with torpedo tubes—approached, hugging the fog-bound coast, and fired four 356-mm torpedoes. One hit the Russian destroyer *Boevoi*, which, although damaged, remained afloat and was towed back to Port Arthur. She took no further part in the war.[79] Two torpedoes stuck *Leitenant Burakov*, which was damaged beyond repair; her crew completed her destruction by blowing her up.

Following a direct order from Tsar Nikolai II, on 10 August 1904 Vitgeft made a second and more determined attempt to break through to Vladivostok. During the resulting Battle of the Yellow Sea, Vitgeft was killed when a pair of shells struck the bridge of his flagship, the battleship *Tsesarevich*. The Russian squadron was thrown into confusion; several ships—*Tsesarevich*, the cruisers *Askold* and *Diana*, and four destroyers—tried to reach Vladivostok, but all wound up interned in neutral ports due to damage and insufficient coal. The cruiser *Novik* also broke away, but was caught by Japanese cruisers and, after being damaged, was destroyed by her crew; the destroyer *Burnyi* ran aground and was blown up by her crew. However, the bulk of the squadron—five battleships, one cruiser and three destroyers—headed back to Port Arthur. During their return, they were subjected to night attacks by Tōgō's flotillas.[80] It was a moonless night, and both the Russians and Japanese frequently lost track of their own vessels and those of the enemy. By now the Russians had learned several lessons: they refrained from using their searchlights, making them difficult to find, and when attacked they turned away from the enemy—a tactic that proved highly effective, especially since the

Japanese persisted in firing torpedoes at long ranges (and, consequently, at low speeds). Individual attacks continued to be the norm, with the boats of a division separating from one another before attacking. No fewer than forty-six torpedo craft—seventeen destroyers and twenty-nine torpedo boats—fired a total of seventy-four torpedoes, scoring only a single hit on the battleship *Poltava*, but thanks to evasive action the torpedo struck at such an acute angle that it failed to explode.[81]

Although the Port Arthur Squadron still constituted a formidable force if used boldly, the ships were stripped of men and light guns to assist in the defenses ashore. The Japanese meanwhile were bombarding the harbor with 11-inch siege howitzers, but at first they had to fire blindly because they could not see over the mountains that ringed the port. That changed on 5 December when they finally captured 203 Metre Hill, which afforded an excellent view of the entire basin. One by one the Russian ships were sunk by the high-angle fire of the howitzers—except for the battleship *Sevastopol*. Under the command of Captain First Rank Nikolai Ottovich von Essen, she was anchored outside the harbor in White Wolf Bay. Between 10 and 15 December, as gales and blinding snowstorms swept the region, *Sevastopol* was subjected to a series of night attacks by Japanese torpedo boats and launches. In addition to her own guns, she was defended by torpedo nets, destroyers, and the nearby shore batteries.[82] During these actions the Japanese launched 124 torpedoes; several were caught in the nets, and only two scored hits: one on the Russian destroyer *Storozhivoi*, which managed to limp into the harbor, and one on *Sevastopol*'s stern. The battleship remained afloat, however, until Port Arthur surrendered on 2 January 1905; at that point von Essen had *Sevastopol* towed into deep water and scuttled her. The Port Arthur Squadron was no more.

Over the course of ten months of war, the Russians had learned some important lessons, including the need to use searchlights judiciously, the effectiveness of evasive action, and the value of defensive fire. But with almost all the crews in captivity or interned, the problem was conveying those lessons to the Second Pacific Squadron, now on its way to the Far East.

TSUSHIMA, 27–28 MAY 1905

The Second Pacific Squadron, commanded by Vice Admiral Zinovii Petrovich Rozhestvenskii, had departed from the Baltic in October 1904.[83] Throughout the journey he received numerous reports from Russian intelligence sources indicating that the Japanese would mount attacks before he reached the Far East, and he therefore took all the standard precautions to foil such attempts: at night, gun crews slept by the rapid-firing guns, which were kept loaded; when anchored, the ships spread torpedo nets; and searchlights were used to illuminate the seaward approaches.[84] Reports and rumors of Japanese torpedo craft in European waters set the stage for the Dogger Bank Incident on the night of 22–23 October 1904, when some of the squadron's ships, convinced that the Japanese torpedo craft were attacking, shot up several British trawlers, sinking one of them. The squadron sailed on without stopping.

During a two-month stay in Madagascar, the squadron carried out exercises in repelling night attacks no fewer than sixty-one times, frequently involving the ships' launches attempting to sneak up on the squadron.[85] Rozhestvenskii also stressed the need to prepare for night actions in his gunnery orders.[86] But the one thing he failed to do was exercise his ships in steaming at night while darkened.

In February 1905 another group of ships was sent to reinforce Rozhestvenskii. Commanded by Rear Admiral Nikolai Ivanovich Nebogatov, it consisted of the Baltic Fleet's leftovers—the elderly battleship *Imperator Nikolai I*, three coast-defense battleships, and an ancient cruiser. While Nebogatov was not a great leader, he was "a man of sound common sense."[87] Like Rozhestvenskii, he took the usual precautions against torpedo attacks, but, unlike Rozhestvenskii, he also trained his ships to steam at night without lights.[88] He also may have gained valuable insights into Japanese torpedo tactics from von Essen, former commander of the battleship *Sevastopol*, who had been released from Japanese captivity by pledging to take no further part in the war. On his way home he was diverted to Suda Bay, Crete, to meet Nebogatov's detachment as it passed through the Mediterranean. He gave a talk to the detachment's officers, and may have shared information about night-fighting techniques, in which he was well-versed.[89] Nebogatov's detachment eventually caught up to Rozhestvenskii in French Indochina and became the Third Division of his squadron, but Nebogatov was given no opportunity to exchange views with his commander.

While the Russians were steaming halfway around the world, the Japanese were busy repairing, refitting and improving their fleet, a process that was completed by February 1905.[90] But the refurbishment of the fleet was not merely material; under Akiyama Saneyuki, now promoted to commander and serving as Tōgō's chief planner, a new battle plan was formulated. Akiyama envisioned subjecting the approaching Russian fleet to night torpedo attacks before the daylight gunnery engagement. The doctrine of firing torpedoes at long range was finally abandoned—British attachés now noted that torpedoes were being set to run at their high-speed, short-range setting.[91] A new secret weapon—linked mines—was developed, consisting of four floating mines fastened along a 100-meter cable; the idea was to drop these strings of mines ahead of the enemy, whose bows would foul the cables and drag the mines into their ships.[92] The Fourth Destroyer Division was especially trained in the use of these mines.

As Rozhestvenskii's squadron approached the Tsushima Strait on 27 May 1905, the seas were so rough that the Japanese torpedo boats had to take shelter in their bases, so Akiyama's plan to whittle the enemy down by preliminary torpedo attacks had to be abandoned. But it made little difference to the fate that awaited the Russians; by sunset, four of their five best battleships had been sunk and most of the other ships had been damaged to some degree. As darkness came on, Admiral Tōgō withdrew northward, positioning his almost-intact fleet for an engagement the next day and leaving the field clear for his torpedo craft to mount a series of night attacks.

Rozhestvenskii's flagship, the battleship *Kniaz Suvorov*, had been wrecked by gunfire during the daylight battle and he had been severely wounded. Barely conscious, he was transferred to a destroyer, leaving Nebogatov in command of the squadron's remnants.[93] His Third Division had survived the day in relatively good shape, although the coast-defense ship *Admiral Ushakov* had been hit by a heavy shell near the waterline forward and was well down by the bow. Nebogatov managed to rally what was left of the squadron, including the battleships *Navarin* and *Sisoi Velikii* and the obsolete armored cruiser *Admiral Nakhimov*. Also steaming nearby was a cruiser detachment under Rear Admiral Enkvist, as well as a handful of destroyers and a few transports (see Map 1.2).

At dusk, Nebogatov saw Japanese torpedo craft coming down from the north, and turned away to the southwest, fearing that they would lay mines in his path.[94] Darkness came on swiftly, since moonrise would not come until about 0140 the following day, and there was still a heavy swell, although it would moderate as the long night wore on. After steaming southwest for a while, Nebogatov decided to turn back to the north, but at about 2000 or

MAP 1.2. SUNSET ATTACK, 27 MAY 1905

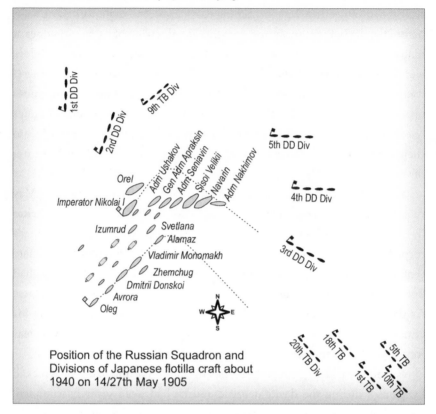

Position of the Russian Squadron and
Divisions of Japanese flotilla craft about
1940 on 14/27th May 1905

somewhat after, as his column was turning through west back to north, the
Japanese torpedo craft caught him.[95]

Several things now happened more or less simultaneously. Nebogatov
increased speed to twelve knots, and several ships that had sustained waterline damage during the day began to lag behind, soon finding themselves
alone. Nebogatov ordered the ships that accompanied him—the old battleship *Imperator Nikolai I*, the new but heavily damaged battleship *Orël*, the
coast-defense battleships *General-Admiral Apraksin* and *Admiral Seniavin*, and
the small cruiser *Izumrud*—not to use their searchlights, and after fighting off
a few attacks without damage, they passed the remainder of the night without
incident.

In a similar fashion, Enkvist, who was still leading his cruisers in a south-
erly direction, increased speed to eighteen knots, and the damaged and older

ships, including the obsolete armored cruiser *Vladimir Monomakh*, were unable to keep up. After beating off a few attacks, Enkvist—now with only his three fastest ships, the protected cruisers *Oleg*, *Avrora*, and *Zhemchug*—tried to turn back to the north several times, but kept running into numerous lights, which he took for the Japanese fleet. In fact, they were probably the fishing flotillas that usually plied their trade in the Korean straits.[96] In the end, he gave up and steamed to Manila, where his ships were interned.

Other ships were not so fortunate. The initial Japanese attacks were made by the First and Second destroyer divisions and the Ninth Torpedo Boat Division, which were all coming down from the north. From the Russian perspective, the onslaught seemed fairly well-organized; Captain First Rank A. A. Rodionov of the armored cruiser *Admiral Nakhimov* described it in the following terms: "The Japanese attacked in divisions of four torpedo craft each. The first attack was carried out at long range, 10–12 cables [2,000–2,400 yards], while the subsequent [attacks] were from 3 cables [600 yards] or less. The divisions at first went on the opposite course from us, launching torpedoes from one side, then they turned onto parallel courses and launched torpedoes from the other side. A division, having fired its torpedoes, disappeared and was replaced by another."[97]

In fact, the attacks were far more confused. In many cases, the boats of the divisions became separated and wound up attacking individually at different times from different directions; some boats lost sight of the enemy entirely, while others never found them at all. The chaos was compounded when subsequent waves of attackers, coming from the east and south, overlapped with previous groups. A Japanese witness described the scene at the height of the attacks: "Numerous destroyers and torpedo boats had surrounded the enemy on all sides, and some could be seen running close up through the glare of the moving searchlights.... With numberless boats rushing like the wind through the darkness, scraping past each other's bows and sides, the danger of collision was very great indeed."[98] So great indeed that *Yūgiri* of the Fifth Destroyer Division rammed *Harusame* of the First Division just as she was going in to attack, while *Akatsuki* of the First Destroyer Division collided with the First Torpedo Boat Division's *No. 69*, which eventually sank as a result. A good many other collisions were only narrowly averted by violent maneuvering.

The attackers were aided, however, by the fact that the Russian stragglers were using their searchlights to scan for approaching enemies. The first victim was *Admiral Nakhimov*, struck well forward in the first wave of attacks. By morning her bulkheads were failing and she made for nearby Tsushima Island, sinking close to shore after being abandoned by her crew.

The battleship *Sisoi Velikii*, which had been damaged forward by gunfire during the day and was now struggling on alone, was hit by a torpedo in her steering-gear compartment at about 2315.[99] Slowly settling forward, by morning she had to steam stern-first, heading for Tsushima also. She surrendered to Japanese auxiliary cruisers, but soon sank despite their efforts to tow her to shallow water.

The armored cruiser *Vladimir Monomakh* was another straggler; at about 2100 she was approached by a destroyer and opened fire, but the latter immediately gave the correct recognition signal—it was the Russian destroyer *Gromkii*, which took station to port of the cruiser. Not long afterward two more torpedo vessels appeared, and again *Monomakh* opened fire, but ceased when one of the newcomers made "some sort of signals," which the cruiser took to be Russian recognition signals.[100] But these boats were Japanese, and one fired a torpedo at close range; *Monomakh* was hit on the starboard side forward. Like *Nakhimov* and *Sisoi*, she headed for Tsushima, and sank near the island the following afternoon.

The battleship *Navarin* is the most interesting—and most tragic—case. She was torpedoed at about 2200, on the port side aft; a low-freeboard ship, she soon found her quarterdeck awash right up to her after turret. After struggling to get a collision mat over the hole, she continued onward at low speed, but at about 0140 she was sighted by the Fourth Destroyer Division. Receiving no reply to their recognition signals, the destroyers laid six strings of linked mines about three hundred yards ahead of *Navarin*.[101] Within minutes they heard a "dull thud" followed by a "loud explosion"—presumably the ship had fouled one of the mine cables.[102] She sank quickly, and although a considerable number of men made it into the water, only three survivors were still alive by the time rescuers arrived the next day.

In contrast, most of the Russian destroyers passed a relatively quiet night. Rozhestvenskii's instructions had given them no scope for independent action during the day, ordering them to protect the squadron's transports or to serve as rescue vessels when ships sank; he assigned no tasks at all for the night after the battle.[103] As noted, *Gromkii* wound up escorting *Monomakh*, stationing herself off *Monomakh*'s port side, but *Monomakh* found *Gromkii* more of a hindrance than a help, since the destroyer got in the way of the cruiser's maneuvers when the enemy attacked.[104] Most of the other destroyers proceeded independently throughout the night without encountering the Japanese.

The day after the battle Nebogatov's ships surrendered to the Combined Fleet, except for *Izumrud*, which broke away but ran aground on the coast of the Kamchatka Peninsula and was blown up by her crew. Some of the other Russian stragglers were sunk by gunfire on 28 May. Rozhestvenskii was captured when his destroyer surrendered. Only three ships reached Vladivostok—the cruiser-yacht *Almaz* and the destroyers *Bravyi* and *Groznyi*; the first two had made their way undiscovered up the Japanese coast, while the latter had done the same along the Korean coast.

During the night twenty Japanese destroyers had fired a total of forty torpedoes, while twenty-four torpedo boats had launched thirty of them.[105] Yet, despite attacking at closer ranges and using the high-speed setting, only four torpedoes found their targets. Although this was a better result than had been achieved in the night attacks on the Port Arthur squadron during its sorties on 23 June and 10 August, it was still an unimpressive performance.

WHAT WENT WRONG?

The Russians had taken the initiative in only one of the actions described here—the destroyer action on 10 March 1904, which was a close-range melee fought primarily with guns; in the other four battles they had been the recipients of Japanese torpedo attacks. Yet, despite the attention paid by the Japanese to torpedo doctrine and despite the intense training, courage, and experience of the destroyer and torpedo boat crews, the results had been minimal. According to one analysis, during the entire war the Japanese had fired 370 torpedoes and scored 17 hits, or about 4.6 percent; another calculation indicates that the hit rate against moving ships was only 2 percent.[106] There were many reasons for this, starting with the fact that Japanese doctrine had initially favored long-range torpedo attacks using low-speed settings, which increased the effects of any mistakes in estimating a target's course and speed. This tendency toward long-range attack was reinforced by defensive fire; even when the gunfire was inaccurate, it served as a deterrent. A report by the Combined Fleet Staff noted that although the boats often reported firing at four hundred to eight hundred meters, in fact torpedoes were often fired from much greater distances, and this was the chief cause of misses.[107] Russian observers support this, stating that the actual firing ranges were usually two to three thousand yards and sometimes more.[108]

Heavy defensive fire also created another difficulty for the attacking boats: when illuminated by searchlights, shell splashes became "brilliantly iridescent fountains of spray" that made it impossible to see their targets.[109]

Another important factor was that even when a division approached as a group, it would separate to attack, making for a series of uncoordinated individual thrusts. Moreover, boats fired single torpedoes, and any miscalculation in the target's course or speed—or any maneuvers that changed these parameters—would cause the torpedo to miss. This meant that evasive action, especially turning away at high speed, was very effective; it simultaneously changed the target's course and reduced its profile, and in some cases the target could even outrun a slow-moving torpedo. The idea of firing spreads of torpedoes to compensate for miscalculations and target maneuvering still lay years in the future.

All of these difficulties were multiplied by darkness. Both attackers and defenders faced challenges in simply maintaining formation; shaded stern lights would disappear if a leading ship turned unexpectedly, throwing her followers into confusion. In several cases this led to ships latching on to the wrong formation, as happened to Sazanami during the initial attack on Port Arthur. She soon discovered her mistake, but Strashnyi was not so fortunate: on the night of 12–13 April 1904, she was sunk when morning revealed that the ships she was following were Japanese.

This brings up another problem—that of identification. In the dark, one destroyer looked much like another, and the first ship making a challenge was always at a disadvantage: she revealed her nationality the moment she began making a recognition signal. Moreover, both sides used red and white lights for night recognition signals, so when Japanese torpedo craft made "some sort of signals," Vladimir Monomakh assumed they were friendly, and paid dearly for her mistake.

Searchlights had proven to be a double-edged sword, illuminating the attackers but also revealing the position of the ship to any others nearby that might otherwise have overlooked her. This should have come as no surprise; it had been realized by the Royal Navy as far back as the 1880s.[110] The Port Arthur Squadron had made profligate use of searchlights early in the war, but by the time of the Battle of the Yellow Sea it used them sparingly. Nebogatov's ships had learned this lesson before Tsushima, but Rozhestvenskii's had not.

These battles, and especially Tsushima, also make clear the limitations of the available means of command and control; with numerous boats acting without a general plan of attack, mutual interference and collisions were inevitable. Even though the Japanese destroyers were equipped with radio, no overall control during action would have been possible because the boats lacked any sense of situational awareness—they often had only the haziest notion of their own positions, their targets, and the results of their attacks, so any reports of the enemy's location, course, and status would have been useless. Moreover, contemporary radio sets were too slow and clumsy to coordinate dozens of vessels in a fast-moving battle.

Human nature also played a role. The excitement of battle often led to errors, and actions tended to be brief and sudden, making fast decision-making crucial. Darkness often gave full play to the imagination—men saw things that were not there, often including convincing details of enemy ships sinking when none had been sunk. In almost every action, both sides overestimated the damage they had inflicted on the other.

A final point is worth noting: even on those infrequent occasions when torpedoes found a target, their relatively small warheads rarely caused damage that was immediately fatal. At Tsushima even the older Russian ships survived for several hours after being struck.

LESSONS APPLIED

Not surprisingly, the war had a profound influence on both the Russian and Japanese navies, not least on the tactics of night actions. After the war and the abortive revolution of 1905, the Russian economy was in a shambles; there was no money to replace the ships lost in the war. Faced with the ongoing expansion of German naval power, the Imperial Russian Navy was forced to adopt a strategy based on mine and torpedo warfare. Under the vigorous leadership of none other than von Essen, the Baltic Fleet's destroyers became expert in tactics based on stealth and surprise.[111] When war came in August 1914 the fleet would launch a campaign of defensive and offensive minelaying, fouling the approaches to German Baltic ports with some of the world's most effective mines.[112]

As for the Imperial Japanese Navy, far from being discouraged by the paltry results of its night torpedo attacks, it committed itself even more deeply to them. With Russia no longer a threat, the United States was taken as the enemy for planning purposes. Unable to match the U.S. Navy's growing dreadnought fleet, the Japanese navy adopted Akiyama's concept of whittling down the enemy by night attacks before the final battleship encounter.[113] In the decades after the Russo-Japanese War it would develop devastating night-fighting techniques—a process that will be described in Chapter 5.

ACKNOWLEDGMENTS

As in so many of my projects, much of the material I used in this chapter was provided through the unfailing generosity of Sergei Vinogradov. Carlos Rivera pointed me to an important source on Japanese officers, while Arseny Malov and Chris Carlson alerted me to sources for Japanese torpedo characteristics. Roberto Romo, Dusha Lin, and Nghia Huynh of the San Francisco Public Library's Government Information Center helped identify the modern names of geographical features. And once again my wife, Jan Torbet, provided invaluable editorial assistance.

CHAPTER 2

TACTICS OF FRUSTRATION
THE GERMAN NAVY AND NIGHT COMBAT, 1914–1916

LEONARD R. HEINZ

Our boats are given deadly weapons against all types of ships
in their torpedoes. You can use them to achieve successes that far
exceed the risks involved. The main task of the torpedo boats
is therefore the torpedo attack.[1]

December 1914. German torpedo boat V155 was headed west in the North Sea, screening the battleships of the fleet, when her lookouts saw ships looming out of the darkness ahead. The torpedo boat's commander, Oberleutnant zur See Hans Carl, ordered slow speed and then stopped his vessel, thinking that the newcomers now slowly steaming by 300 meters away were friends. But their response to a challenge made by signal lamp was gibberish, so Carl had his ship flash her new colored recognition lights.

> No answer! *V155* maximum power! 0625—radio message sent: "Enemy destroyers in sight!" Distance now 2,000 meters, *V155* goes to half speed to reduce her smoke. Turns to an easterly course. The third boat searches for *V155* with her searchlight without success and her light fades. But an enemy boat opens fire. . . . The enemy forms a semicircle with an advanced right wing and runs up quickly, opening fire from all directions; *V155* replies with her rear gun, range setting 2,000 meters, about 5 rounds fired. 0648—radio message sent: "Chased by destroyers in 154β." Flank speed.[2]

Carl was lucky; despite having been pursued by seven British destroyers, *V155* escaped unscathed, and even knocked two opponents out of the fight with her 8.8-cm guns. But this was not the sort of war that the German torpedo boats were meant to fight.[3]

THE STRATEGIC NEED

Germany's naval leaders faced a surfeit of enemies—not just the French navy and the Russian Baltic Fleet, but the British navy, the largest in the world. Accordingly, the Germans sought means to help tip the balance in their favor as much as possible. They built capital ships capable of absorbing great punishment and paid scrupulous attention to damage-control techniques. They planned to fight close to their home ports, so that their damaged ships would have the best chance of reaching safety and that damaged enemy ships would have the greatest chance of sinking. And most importantly, they planned to pit their light forces against British battleships and battle cruisers and to focus their total strength against isolated British forces. The Germans called the war of light forces *Kleinkrieg*. Applied to land warfare, Kleinkrieg simply meant guerilla warfare; at sea, it described using mine and torpedo warfare in order to wear the enemy down before a clash of capital ships that was expected to be decisive.

In the first years following the Russo-Japanese War, naval thinkers saw torpedo boats as the preeminent practitioners of torpedo warfare, with night fighting as their forte. Larger warships and submarines also carried torpedoes, but submarines then were unproven platforms, which the Imperial German Navy—the *Kaiserliche Marine*—adopted only grudgingly in 1906. And torpedoes carried by the big ships so far had proven ineffective in combat. Night operations were seen as necessary, for although both the size and capabilities of torpedo boats grew markedly after 1905, the defenses against them also increased in both scale and scope. Big ships acquired heavier batteries of anti-torpedo boat guns, and navies built more torpedo boat destroyers. If the boats were to get close enough to make effective attacks, the cover of night was still necessary. German thinking conformed to this. The Germans painted their torpedo boats black and dubbed them the *Schwarze Schwar* ("Black Horde").[4]

Much longer-range torpedoes altered this calculus. In 1908 the naval periodical *Marine Rundschau* heralded the new era with a report of a 6,000-yard torpedo developed by the Whitehead company, amplified in early 1909 with

PHOTO 2.1. The *Schwarze Schwar.* A half-flotilla of five German torpedo boats painted black and steaming in a characteristically tight wedge formation. The closest boat is *S138*, which dates this photograph to 1907 or later. The photo shows the smoke that coal burners could produce, even at moderate speeds. *U.S. Naval History and Heritage Command, NH 45406*

a report of a 21-inch British weapon that could run to 7,000 yards at 31 knots.[5] Proponents argued that these new weapons, which bid fair to rival guns in their effective ranges, could broaden the scope of torpedo attacks to daytime actions. Once the German torpedo boat force got long-range torpedoes, its officers began to explore a daylight role for the boats. A follow-up article in *Marine Rundschau*, in 1910, still called night attack the most important mission of the torpedo boat force.[6] By 1914, however, the German tactical manual for the torpedo boats devoted as much space to day attacks as to night attacks.[7]

The Kaiserliche Marine plan for 1914 assumed that British ships would conduct a close blockade of the German bases in the North Sea, and that strikes on those ships would force the commitment (and hoped-for destruction) of progressively heavier Royal Navy forces. Torpedo boats figured prominently now: they were tasked with night attacks, both independently and in support of the main battle fleet (the *Hochseeflotte*); with daylight attacks in concert with the Hochseeflotte; and patrolling to keep marauding British ships and submarines at bay. Even the idea of conducting raids against the British coast if the Royal Navy proved reluctant to engage lurked in the wings.[8]

TOOLS AND TECHNOLOGY

The Kaiserliche Marine manual for torpedo boat tactics summed up the desirable attributes of a torpedo boat:

- ▶ Powerful torpedo armament
- ▶ High speed
- ▶ Maneuverability
- ▶ Seaworthiness
- ▶ Smallest possible size[9]

An August 1910 article in *Marine Rundschau* echoed those five criteria. It also considered gun power and various measures to protect crew, machinery, and magazines, but it dismissed them as being of lesser importance. Radius of action did not make the list. The author acknowledged that a good design required a reasonable compromise among the five critical factors, but stressed that the boats needed a heavy torpedo armament to accomplish their main mission—attacks at night.[10]

Central to a powerful torpedo armament were prewar enhancements in torpedo performance—the result of improved propulsion technology, better warhead explosives, and bigger torpedoes. Seeking gains in speed and range, the German navy supplemented its 45-cm torpedoes with a 50-cm model adopted in 1910. The initial version gave a range of 5,000 meters, increased to 9,000 meters by the start of the war. The big torpedo used a more potent explosive (a densely packed mix of trinitrotoluene and hexanitrodiphenylamine), in a warhead weighing up to 200 kilograms.[11] Given the difficulties of judging a target's course and speed at night, the navy concluded that the longest ranges of the new torpedoes could only be exploited by daylight attacks.[12] Although the latest and best models first went to battleships and battle cruisers, the newest torpedo boats received improved weapons with enough range to give them a role in daytime battle—a role that was extensively tested in prewar maneuvers.[13] But when naval planners combined this new task with night-attack and routine patrol duties, the torpedo boat force risked becoming a jack of all trades and a master of none.

The first German torpedo boats with 50-cm torpedoes carried five of the new weapons—four ready in deck-mounted tubes, plus one reload.[14] Contemporary designs in other navies typically had fewer tubes, lighter torpedoes, or both. German oceanic boats launched after 1913 mounted six

50-cm tubes—two single tubes forward, with one firing to port and one to starboard, plus twin mounts amidships and aft.[15] Each twin mount carried its tubes at a twelve-degree angle.[16] Mounts could be set at fixed angles or rotated, although the forward tubes were hard to work in a seaway and usually were left fixed in position.[17] At least in the latter half of the war, boats carried their main torpedo directors port and starboard on the bridge with cruder directors at each mount. The bridge directors could fire all tubes.[18] Although a six-tube ship could theoretically fire a five-torpedo broadside, the angled twin mounts—along with the admonition that every torpedo fired at night should be a hit—resulted in a boat firing only one or two at a time. Even in the daytime, the maximum salvo that the boats actually fired seldom exceeded three torpedoes.[19]

The prewar Imperial German Navy differed from contemporary navies in that it never commissioned any torpedo boat destroyers—with a single exception, purchased from a British yard.[20] Other navies—particularly those of Britain, Russia, and Japan—saw destroyers as multirole vessels that balanced their torpedo armament with a gun battery powerful enough to bolster defenses against enemy torpedo boat attacks. The Germans resisted building these types in favor of torpedo boats that carried heavier torpedo armament. German naval thinkers argued that destroyers were poor gun platforms and that the Russo-Japanese War had demonstrated the ineffectiveness of destroyer guns. Some did fear that Royal Navy destroyers and light cruisers would stymie German torpedo attacks against the British battleline, but others dismissed the growth in the weight of gun batteries with the argument that larger torpedo boats could resist the power of heavier guns.[21] German tactical doctrine relied on cruisers to support torpedo boats that were faced with enemy screening forces, while the boats themselves were to (1) avoid contact if possible, (2) attempt to engage the enemy with only a portion of the torpedo force if not, and (3) accept a full-scale action with opposing light forces only as a last resort.[22] As tactical doctrine made explicit, Germany needed its torpedo boats to carry weapons for offensive action, not defensive reaction.

In the quest for speed, German torpedo boat designs incorporated two important advances in marine propulsion—turbines and oil fuel. Turbines brought both advantages and disadvantages. The chief disadvantage was a loss of range relative to the highly developed piston engines of the day. But, given the greater speeds that the new design offered—together with less vibration, simpler operation and maintenance, personnel savings, and lighter weight—the navy readily accepted the turbine's range limitations. Most of its boats had ranges at combat cruising speeds of about a thousand miles, and the service considered this good enough.[23] Oil fuel brought considerable advantages over coal: a higher caloric content; more constant feeding, rather than having to rely on stokers to bring coal from distant, awkwardly located bunkers; and, most important for night fighting, less chance of a boat revealing itself. A coal-fired boat running at high speed left billows of smoke and showers of sparks in its wake, potentially alerting enemies to its presence. German fleet orders noted that coal-fueled boats could cruise at fifteen knots for only three hours before smoking heavily, but this meant letting their fires burn down to the point where they no longer could develop high speed for attacks.[24] Oil-fired boats could smoke and "torch" as well, but the risk was far less.[25] The main worry for the Germans was whether adequate supplies of oil would be available in wartime. The navy commissioned its first class of turbine-powered boats in 1909, but boats fueled by oil alone did not appear until 1914.[26] The British were a year ahead in deploying a class of turbine-powered destroyers. Since these were also oil fired, the British were six years ahead of the Germans in adopting that technology.

PHOTO 2.2. Size matters: coastal torpedo boats moored in front of a flotilla of *S90*-class oceanic boats. *S90*, commissioned in 1899, had more than double the displacement of *S63*. By the start of the war displacements would double yet again. Night-attack tactics did not keep pace with the increased size of the boats and the increased range of their weapons. *U.S. Naval History and Heritage Command, NH 47696*

TABLE 2.1. GERMAN TORPEDO BOAT DEVELOPMENT

YEAR	SHIP	DISPLACEMENT	SPEED	GUNS	TORPEDO TUBES	PROPULSION
1906	G132	544 tons	28 knots	3 × 5-cm/40	3 × 45-cm	Piston, coal
1910	S176	781 tons	32 knots	2 × 8.8-cm/30	4 × 50-cm	Turbine, coal and oil
1914	V25	975 tons	33.5 knots	3 × 8.8-cm/45	6 × 50-cm	Turbine, oil
1914	Lydiard	1150 tons	29 knots	3 × 4-inch/45	4 × 21-inch	Turbine, oil

Displacement is full load displacement in long tons. Actual speed and range varied widely depending on sea state and the condition of the ship's machinery. *Lydiard* of British *Laforey* class shown for comparison.

Sources: Gröner; Gardiner, ed., *Conway's 1906–1921*; Norman Friedman, *British Destroyers: From the Earliest Days to the Second World War* (Annapolis, MD: Naval Institute Press, 2009).

Larger boats accommodated larger torpedo batteries, but they sparked concerns as well. By 1909 the size of new boats prompted protests from the fleet, where officers were worried that the decreasing maneuverability and increasing silhouettes of the craft might hamper them in combat. The result was a 14-percent reduction in displacement from the 1910 *G192* class to the 1911 successor, *V1* class—from 797 tons to 686 tons at full load—but the *V1*s struggled with seakeeping and were soon dubbed "Lans' cripples" (*Lanskrüppel*), named after Konteradmiral Wilhelm Lans, the unfortunate Torpedo Inspectorate director who oversaw their design.[27] Larger boats quickly returned, but German torpedo boats remained smaller than contemporary British destroyers. Maneuverability and a low silhouette continued to be priorities.[28]

The Kaiserliche Marine worked hard to develop technologies to support night fighting. It particularly focused on searchlights, night signaling, and night-recognition procedures. The navy had successes in searchlight technology and telling friend from foe, but it failed to develop signaling technology that allowed it to fully exploit the offensive potential of its boats. German searchlights were powerful and used lightproof iris shutters that allowed them to be deployed or shut off instantly without waiting for a light filament to start or stop glowing. Searchlights in larger ships were tied to gunnery control systems, enabling crews to lay the lights and guns on the same target before the lights illuminated it.[29] The Germans did not use searchlights to search for the enemy. Ships relied on visual sighting, unaided by illumination, generally not using their searchlights until they were about to engage a sighted target.[30] Searchlights had other uses as well, such as dazzling enemy gunners and fire-control parties.[31] As star shells came into use, the navy preferred them to searchlights, particularly when a ship was engaging land targets or other vessels at longer ranges.

The navy relied on a wide variety of devices and techniques for nighttime signaling. One device signaled course changes by colored lights, and another indicated speed changes by lighted geometric patterns. A "Night Signaling Apparatus" used white and red lamps in various configurations to transmit the number groups of the German naval code. Other means included signals flashed from lamps and searchlights, steam whistle signals, colored rockets, and even—for torpedo boats—megaphones.[32] The methods that the Germans adopted worked reasonably well for normal cruising, but were far less useful for tactical signaling in night battle. The torpedo boat tactics manual prohibited light signals when torpedo boats were steaming without lights, but the alternatives—sound signals—were limited in range and in the information they could carry.[33] While the oceangoing torpedo boats all had radios, the devices were used for reporting sightings and not for the tactical command of small formations. Radio duties taxed the small crews of the boats, while vibrations when the boats exceeded fifteen knots made reception of radioed dots and dashes unreliable.[34] The German concept of torpedo boat attacks was for boats to pursue rather than ambush their targets so that their radios became useless just as the boats were steaming into action." Voice radio might have offered a potential solution to nighttime command and control problems, but despite the existence of a vibrant domestic radio industry, the navy's own enthusiastic use of excellent radio telegraph sets, and even a 1907 trial of voice radio, no workable voice sets emerged.[35]

The navy tackled the problem of telling friend from foe at night by equipping its ships with a series of red, white, and green lights set on their masts. A challenged ship would flash a prearranged configuration of lights to identify itself. The system resembled the two-color Night Signaling Apparatus, which could also be used for night recognition. The navy introduced the three-color system just before the start of the war.[36] The system was quick to use and (as the British acknowledged) hard to mimic—two important qualities for any night-recognition system.[37] The Germans tried to keep this system secret, with crews instructed to rig it only at dusk and strike it at dawn so that it would not be "seen by unauthorized persons."[38] The navy did not rely entirely on its recognition system. For night cruising, torpedo boats were told to keep apart from capital ships and the German signal regulations cautioned torpedo boat captains that approaching a friendly capital ship at night risked gunfire in response.[39]

ORGANIZATION, TRAINING, AND DOCTRINE

The Kaiserliche Marine typically built torpedo boats in groups of twelve. Eleven boats formed a flotilla, with the twelfth boat unmanned and in reserve. Where possible, each flotilla contained ships of the same class. Each flotilla had a leader boat and two five-boat half-flotillas; each half-flotilla had its own leader boat and two two-boat groups. A korvettenkapitän usually commanded each flotilla and half-flotilla while a kapitänleutnant or an oberleutnant zur see skippered each boat. Tactically, flotillas also could deploy in three groups of three boats, each led by the flotilla leader or a half-flotilla leader, plus a two-boat group.[40]

The torpedo forces expanded significantly before the war. In 1909 the Germans deployed five flotillas.[41] In 1910 a sixth flotilla began training.[42] By May 1914 the navy had seven flotillas—two active, two in reserve, and three training.[43] An eighth flotilla was formed at the start of the war, although with a hodgepodge of boats.[44] The expansion in the torpedo boat arm coincided with burgeoning manning requirements for cruisers and capital ships, and the navy struggled to fill its billets. This was felt keenly in the torpedo boats, where their small crews were thinly stretched even when at their full complements.[45]

Petty officers and seamen selected into the torpedo branch served there throughout their time in the ranks. Crewmen were drafted for three years of active service, although petty officers (the backbone of any navy) served longer. Commissioned officers with torpedo training served in other commands, but all had to maintain a torpedo training qualification to serve in the boats.[46] A review of officer assignments in the prewar years shows that these officers often served side-by-side both ashore and afloat and so formed a closely knit group. A flotilla commander typically would have skippered a boat and commanded a half-flotilla before assuming a flotilla command. Shore assignments with the Torpedo Inspectorate (responsible for the boats, their equipment, and the training of officers and crew) also were common. In sum, German torpedo boat officers were thoroughly inculcated in their craft and doctrine.

Training and maneuvers stressed night operations, including night torpedo firing, shadowing enemy ships, and participating in attacks by large groups of torpedo boats.[47] The 1914 program for fleet maneuvers included ten night exercises, three of which involved the Hochseeflotte's full complement of seventy-seven torpedo boats, plus a daylight tactical exercise that extended into the night and a multiday strategic exercise that anticipated night torpedo attacks.[48] Three of the night exercises included possible torpedo firing—which was significant, given the cost of torpedoes and the difficulties of retrieving them in the dark. The navy paid an additional price for its night maneuvers: seven torpedo boats were involved in nighttime collisions between 1910 and 1913; three of them sank.[49]

The Germans maneuvered their boats in compact formations.[50] Given the short range of their favored tactical night-signaling system—the megaphone—concentration made sense despite the dangers involved when the units carried out rapid maneuvers. Ideally each group of two flotillas would have a cruiser assigned to it, although in practice other duties for the cruisers kept this ideal from being achieved. The assigned cruisers were to provide navigational support, improved platforms for spotting the enemy, and superior radio facilities. Tactical regulations directed them to help torpedo boats break through enemy screens and to support attacks directly with guns, torpedoes, and searchlights.[51]

Despite the increasing range and speed of torpedoes, night attacks were to be made at close range. The German manual for torpedo boat tactics called for attacks on large ships at ranges of three hundred to four hundred meters. At less than three hundred meters, the torpedoes would not have reached their set depths. Beyond six hundred meters "a successful shot against an enemy underway is doubtful, owing to the uncertainty of accurately estimating his speed and course." For actions against destroyers, where torpedoes did not have to run deep for maximum effect, the manual gave one hundred meters as the best range. It admonished: "A hit must be obtained with each torpedo fired at night. So close with the enemy!" Tactical exercises anticipated that attacking torpedo boats might be sighted while fifteen hundred meters from their targets, but still required the boats to close to short range before firing.[52]

German tactical manuals emphasized surprise in night attacks, but also discussed the use of searchlights as an attack aid. Cruisers in contact with the enemy were directed to sweep targets where necessary to show them to attacking boats and to dazzle the enemy when able to do so without illuminating the attack.[53] Torpedo boats were directed to use searchlights and white signal rockets as needed to determine the course and speed of sighted targets and to use searchlights to dazzle them.[54] Thus, while German doctrine paid lip-service to stealthy attacks, it was more attuned to attacks on blinded or distracted targets than on unsuspecting ones. Given that boats were expected to attack at short ranges despite their increased size, this may well have been the result of practical experience in maneuvers, which showed that attacks at the ranges urged in the torpedo boat tactical manual were unlikely to go unnoticed.[55] Mass attacks were encouraged, to compensate for loss of surprise. "The aim must therefore be to attack the enemy with as many torpedo boats as possible, simultaneously or in quick succession and from different sides."[56] Torpedo boats were not to evade or abort just because they were spotted; the *Torpedobootstaktik* manual directed that they must attack "regardless of the consequences to the boats."[57]

READY FOR WAR?

Arguably, the German torpedo boat force was prepared to fight the war for which it had trained. But was it ready for the war that it was about to fight? There were reasons for doubt on both strategic and tactical levels.

For years, the Kaiserliche Marine had wargamed a possible conflict with Great Britain. The games experimented with various strategies, ranging from attacks on the Royal Navy at the start of the war to the withdrawal of the German navy into the Baltic, and the results shaped the role of the torpedo forces—particularly in the decisions that the boats should accompany the Hochseeflotte into any engagement and that unsupported raids by the boats on the English coast were likely to be more perilous than profitable. But the wargames provided no answer to a distant British blockade that deprived the torpedo forces of targets close to the German coast.[58] As Vizeadmiral August von Heeringen, the head of the naval staff, said in 1912, such a blockade by the Royal Navy would make the role of the German battle fleet "a very sad one."[59]

As noted, the German navy saw tremendous expansion in the lead-up to World War I. And while the torpedo forces were acknowledged to be an important part of the navy, the high command ultimately paid much more attention to battleships and battle cruisers. Torpedo boat crews trained extensively, but night-attack training had not kept pace with changes in weapons, platforms, or targets. The tactics prescribed by the *Torpedobootstaktik*—each boat launching a few torpedoes from within six hundred meters—were geared for a mass attack on a lumbering target dazzled by searchlights and beleaguered from all sides.[60] Doctrine accepted that the boats could well be spotted before attacking and made no attempt to exploit the improved performance of torpedoes. Nor was it well-suited for attacks on nimbler opponents. And beyond that, the tactics manual did not resolve the conflict inherent between its injunction that no torpedo should miss at night and its assertion that mass attacks were needed to be effective.

The insistence on close-range attacks may have been borne of the frustrations that the Japanese experienced with long-range torpedo attacks in the Russo-Japanese War. But conditions had changed since then. Torpedoes had gotten faster and now were more likely to hit at longer ranges; at the same time, torpedo boats had grown. Larger boats would be seen at longer ranges, but they also carried more torpedoes. Despite this, there is no hint in *Torpedobootstaktik* that techniques for night attacks at ranges longer than six hundred meters had been considered. Contrast this with the assessment by Capt. William S. Sims, commanding the U.S. Navy's Atlantic Fleet torpedo flotilla, who in 1915 characterized night firings from fifteen hundred yards as "short range."[61] (See Chapter 6.) Sims was not speaking theoretically; in a night attack against moving targets, his flotilla scored at least eleven and likely thirteen hits from eighteen torpedoes fired.[62] The difference in approach cannot be explained by differences in torpedo performance. Rather, German naval doctrine for night attacks had not evolved from the era of small torpedo boats and short-range torpedoes.

German naval officers clearly were aware of the new tactical possibilities offered by better-performing torpedoes, but they focused their thinking on daylight cooperation between torpedo boats and the battleline rather than on night attacks. This can be seen in the 1914 fleet exercises, where many of the large daylight exercises specifically asked participants to comment on the effectiveness of torpedo boat attacks.[63] In contrast, the ten night exercises focused far more on the role of light cruisers in supporting or frustrating attacks. Assessments of torpedo attack effectiveness were requested for only three night exercises. In at least one of the live-fire exercises, anchored ships served as the targets—a test more attuned to recovering fired torpedoes than to testing the skills of the firers. It was natural that the new uses for boats in day actions would attract the most attention, but it also was natural that the emphasis on daylight operations drew attention away from the process of adapting nocturnal operations to new conditions.

The 1914 exercises also highlighted another unsolved problem for German tactical doctrine—finding the enemy at night. None of the exercises called for a night search to find the enemy. All assumed that contact was being maintained as darkness fell, perhaps reflecting a recognition that finding the enemy at night would be difficult. The British had experienced this in their 1913 maneuvers, and it is likely that the Germans had suffered the same frustrations.[64] Likely reflecting the British experience, Royal Navy commander K. G. B. Dewar noted in 1913 that the chances that torpedo craft could locate a fleet in open water at night were "extremely small." Based on this, Dewar felt that "the danger of night torpedo attack has in fact been unduly exaggerated."[65]

On the eve of war the Kaiserliche Marine had excellent equipment and well-trained, if thinly stretched, crews. But the years before the war had seen German planners frequently frustrated when trying to give the navy, as Vizeadmiral von Heeringen said, a "militarily useful chance against England."[66] Further, the focus on capital ships and daytime actions had distracted naval tacticians from closely examining the issues raised by night combat waged with the latest ships and weapons.

BEST LAID PLANS, EARLY STRUGGLES

A Kaiser order of 30 July 1914 summarized the Imperial German Navy's strategy for the coming war. British units blockading the German Bight would be attacked, while minelayers and submarines would operate against the British coast. The battleships and battle cruisers of the Hochseeflotte would engage their British counterparts once attrition tipped the balance in their favor. If a "favorable opportunity" to fall on a detached portion of the British fleet were to present itself sooner, then the Hochseeflotte would exploit that.[67]

The first days of the war forced the Kaiserliche Marine to acknowledge that its war plan had failed fundamentally. No blockading ships appeared off the German ports. Nor did isolated detachments of the Royal Navy present themselves for defeat in detail. Probes by submarines and light cruisers found little to report. The British navy had instituted the strategy that prewar German planners had feared most—a distant blockade with its main forces in Scottish anchorages, well beyond the reach of German torpedo boats. Faced with this, by mid-August the German command had reconciled itself to a naval war that for the time being would consist mainly of minelaying and submarine missions into the North Sea.[68] The one offensive sweep that the torpedo boats undertook, when two light cruisers and VI Flotilla ranged as far as Dogger Bank on 21–22 August 1914, netted only a few fishing trawlers. A British daylight raid into the Heligoland Bight on 28 August provided the first combat for the North Sea torpedo boats, and for the Germans it was a dismal affair that saw the Hochseeflotte's heavy ships trapped in harbor by unfavorable tides and its light forces sliced up by Royal Navy destroyers, light cruisers, and battle cruisers. The British sank torpedo boat *V187* and three light cruisers. The first night attack by a German torpedo boat would not take place until October, and that was on the other side of the world.

On 17 October 1914 torpedo boat *S90* slipped out of the German base at Tsingtao (today Qingdao), China, to attack blockading Anglo-Japanese forces. She sank the old Japanese cruiser-minelayer *Takachiho* with heavy loss of life, but that proved to be the only German success for the classic operational concept of torpedo boats ambushing close blockading forces.[69] The December fight between *V155* and seven British destroyers has been described in the introduction to this chapter. *V155* had been forced into a scouting and screening role by the lack of light cruisers that ideally would have been employed, and her captain found himself in a fight where, once surprise was lost, he had little hope of using his boat's torpedoes effectively.[70] All credit to *V155*'s crew for standing off a greatly superior force, but this defensive fight was far from the raison d'être of the German torpedo boats.

PHOTO 2.3. Torpedo boat *V156*. A sister to *V155*, which battled British destroyers in December 1914. The photo shows her forward torpedo tubes in the break between her forecastle and her bridge. These forward tubes, a common feature of German torpedo boats, were hard to work in a seaway and were generally left in a fixed position and fired from the bridge. In this picture, *V156*'s two 8.8-cm guns had not yet been mounted. *U.S. Naval History and Heritage Command, NH 65793*

On 16–17 August 1915, *V99* and *V100* mounted a night raid into the Gulf of Riga to attack the Russian battleship *Slava*. The two boats managed to creep past the minefields guarding the gulf, but found nothing but trouble inside. They were first harassed by destroyers patrolling the gulf entrance, then by two more destroyers in the gulf. The Germans failed to find *Slava*, and lost *V99* to mines and Russian guns while crossing the minefields on the way out. Such were the difficulties of night searching, particularly in the face of active opposition.[71]

The navy's next torpedo action took place in the North Sea shortly after sundown on 17 August 1915. It pitted five boats of II Flotilla, led by Korvettenkapitän Heinrich Schuur, against a British force of eight destroyers and a minelayer. Schuur had entered the Imperial German Navy in 1893 as an officer cadet. His short stature might have denied him a naval career—he was barely more than five feet tall—but his mother had connections in the service that paved the way for her son's enrollment. Schuur's first assignments were typical of the time, even for officers who later specialized in torpedo warfare—postings to larger ships and various shore billets without a focus on torpedoes. That changed in 1902, when he assumed command of torpedo boat *S105*. Schuur spent three years commanding torpedo boats, being promoted to kapitänleutnant in that time, and then embarked on a two-year assignment as executive officer in gunboat *Iltis*, stationed in Tsingtao. The foreign detour was the navy's response to Schuur's request for permission to marry. He returned to the torpedo forces in 1908, leading half-flotillas for three years, then spending a year at the Torpedo Inspectorate as a korvettenkapitän before assuming command of II Flotilla in 1912. Schuur's two half-flotilla commanders, kapitänleutnants Heinrich Boest and Adolf Dithmar, had not served in the half-flotillas under Schuur's command, but both skippered boats in the same half-flotilla during the time that Schuur was a half-flotilla leader.[72]

Schuur appeared to have decisive advantages over the British; his boats were new oil-burners, their leaders were seasoned, and the enemy had been spotted silhouetted against the western afterglow at least eight thousand meters away while the Germans were to the east, nearly invisible against a bank of storm clouds. Despite these advantages, II Flotilla managed only three torpedo shots from a single boat. Just one torpedo connected, blowing the bow off a British destroyer that nonetheless managed to limp back into port. The German official history blamed the paucity of results on a bright night that prevented a close-in attack, the lights of a German fishing fleet threatening to silhouette the torpedo boats, and smoke from one of the boats, but Schuur had maneuvered for twenty-two minutes and closed to within 3,200 meters of the British before being spotted. Dithmar, in *G103*, the fourth ship in line, later reported that he could have launched torpedoes at long range after the first seven minutes of maneuvering, but held his fire when he received no orders to shoot.[73] Clearly something was amiss—an inability to coordinate an attack, a reluctance to expend torpedoes in long-range fire, or some uncertainty over whether the British ships were profitable targets. Torpedoes were expensive weapons, and so no navy endorsed their profligate use, but German doctrine stressed parsimony. Doctrine also focused on reserving torpedoes for use against enemy capital ships, leaving ambiguous when it was appropriate to expend them on light forces. These shortcomings were cloaked, moreover, by Schuur reporting the sinking of a light cruiser and a destroyer. "Strike effectively first!" would in time be the watchword of night combat, but for the Germans that realization had not yet dawned.

TO DOGGER BANK

When Vizeadmiral Reinhard Scheer became head of the Hochseeflotte in January 1916, he brought with him a more aggressive attitude. Seeking ways to strike at Great Britain, Scheer advocated a program of submarine and minelaying Kleinkrieg, as well as nocturnal Zeppelin raids on British cities and torpedo boat sweeps into the North Sea. The Zeppelin raids and torpedo boat sweeps were to be coordinated, based on the expectation that the Zeppelins would draw Royal Navy forces into the North Sea to form spotting lines and antiaircraft barrages. Thus, the airships would give the boats targets and the boats would protect the airships. The net effect would be to expose British light forces to defeat in detail, just as if they were maintaining a blockade of the German Bight.[74]

Plans for a combined Zeppelin and torpedo boat operation faltered in early February due to bad weather and then died as the moon waxed, precluding Zeppelin operations, but the Hochseeflotte staff persevered with the torpedo boat sweep.[75] The operation began on 10 February, with II, VI, and IX flotillas—a total of twenty-five boats—deployed in eight groups searching westward. These were all oil-fueled vessels, with higher sustained speeds and lower nighttime visibility than their coal- or mixed-fuel sisters. The search line was backed by light cruiser *Pillau*, in which rode the commanding officer,

MAP 2.1. DOGGER BANK RAID, 10–11 FEBRUARY 1916

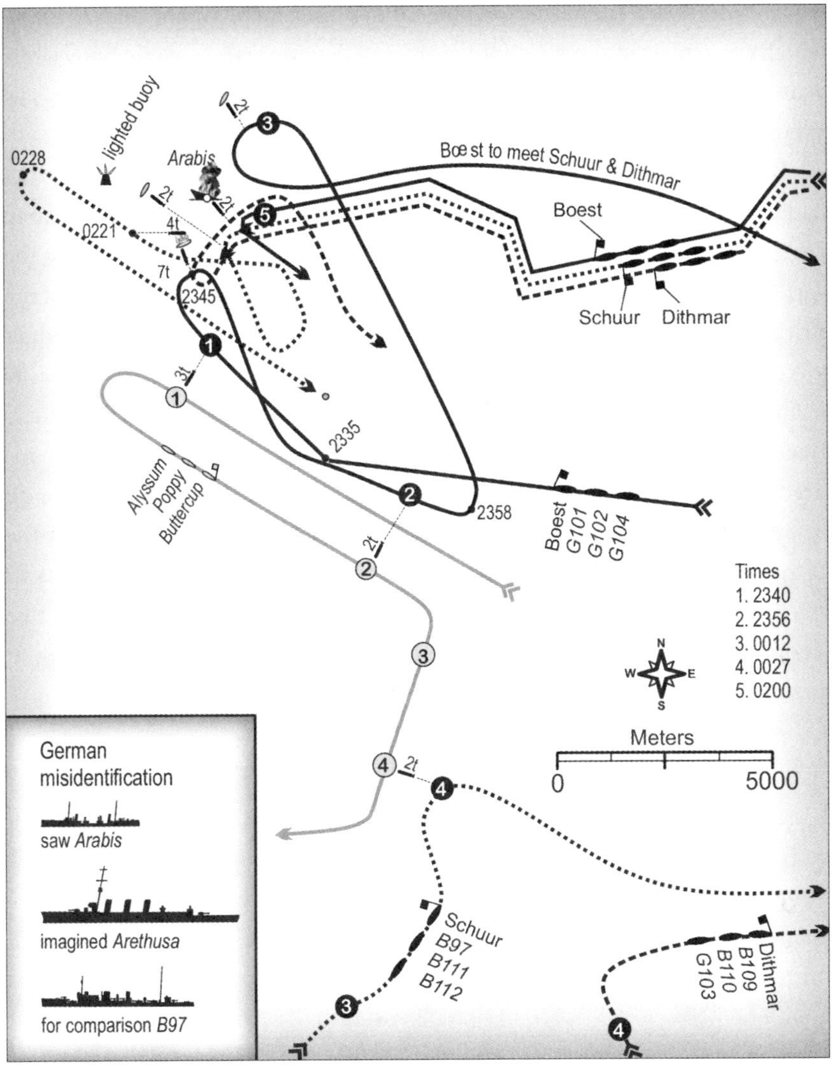

lighted buoy

0228

Arabis

0221

2345

Boest to meet Schuur & Dithmar

Boest

Schuur Dithmar

2335

2358

Alyssum
Poppy
Buttercup

Boest
G101
G102
G104

Times
1. 2340
2. 2356
3. 0012
4. 0027
5. 0200

N
W E
S

Meters
0   5000

German
misidentification

saw Arabis

imagined Arethusa

for comparison B97

Schuur
B97
B111
B112

Dithmar
B109
B110
G103

Kapitän zur See Johannes Hartog. The boats would reach the limit of their search at midnight—about two hundred miles from the Elbe, at the eastern border of the Dogger Bank. As the boats approached their turnaround point, the first-quarter moon was dipping to the west, with three thousand meters' estimated visibility in that direction but less to the east.

At 2310 the lead boat of the northernmost group (three boats of II Flotilla, led by Boest, the commander of 3rd Half-Flotilla, in *G101*) saw to the west the sternlights of two ships followed by a third darkened ship, all heading north-west. These were the minesweeping sloops *Buttercup*, *Poppy*, and *Alyssum*, which were steaming on a circuit toward and then away from a light buoy while waiting for daylight to renew their sweeping. Soon after the German crews saw a fourth sloop, *Arabis*, which had been ordered to stand by the light buoy.

Boest first thought that he might have sighted a German formation that had wandered off course, but the burning stern lights weighed against that. He then concluded that the ships were *Arethusa*-class light cruisers—just the sort of quarry he expected. He ran his three boats up the starboard side of the British formation and fired three torpedoes from twelve to fifteen hundred meters, only to see the British reverse course to port as they turned away from the buoy. All torpedoes missed, but the British did not notice the attack. Boest then led his group to starboard, directly toward *Arabis*. The Germans fired seven torpedoes at her as they swung through their turn and bore off in pursuit of the other three sloops. Again none hit. The boats caught the British formation at 2356, but this time an alert watch officer in *Buttercup* saw the flashes of torpedoes being fired. He turned *Buttercup* away, the other two sloops did likewise, and the two torpedoes that were launched in the attack, at a range of eighteen hundred meters, missed once again. The Germans had now

TABLE 2.2. SHIP CLASSES ENGAGED AND IMAGINED, DOGGER BANK RAID

SHIP	DISPLACEMENT	LENGTH	DRAFT	SPEED	GUNS	TORPEDO TUBES	PROPULSION
British:							
Arabis	1250 tons	268 feet	11 feet	16 knots	2 × 4.7-inch	None	Piston, coal
Arethusa	4400 tons	436 feet	13 feet, 5 inches	28.5 knots	2 × 6-inch/45, 6 × 4-inch/45	4 × 21-inch	Turbine, oil
German:							
B97	1814 tons	321 feet, 6 inches	11 feet, 3 inches	35.5 knots	4 × 8.8-cm/45 or 10.5-cm/45	6 × 50-cm	Turbine, oil
G101	1707 tons	312 feet, 8 inches	12 feet, 2 inches	33.7 knots	4 × 8.8-cm/45 or 10.5-cm/45	6 × 50-cm	Turbine, oil

Displacement is full load displacement in long tons for except for *Arabis*, where it is normal displacement. The *B97* and *G101* classes were far larger than the norm for German boats, the former being built around machinery ordered for Russian destroyers while the latter ships had been seized while under construction for Argentina; the Germans accordingly referred to II Flotilla as the "destroyer flotilla" (*Zerstörer-flottille*).* These ships had their 8.8-cm guns replaced with 10.5-cm guns beginning in January 1916.

*F. Ruge, *SM Torpedo Boat B110* (Windsor, Canada: Profile Publications, 1973), 50; ADM 137/2085, *Information Derived from Examination of Prisoners from the Dutch S. S. "Oldambt" on 2nd November 1916*, 5.

expended twelve of their eighteen ready torpedoes without result, while the British formation zigzagged away. Its commander, Lieutenant Commander Ronald Mayne, had not seen or been told of the flashes seen by *Buttercup*'s watch officer and thought he was evading a submarine attack.

Boest now realized that he had made no report of his contact. He corrected that omission at midnight, although his dead-reckoning put his position five miles too far east and he failed to give an enemy course. With the three supposed cruisers disappearing to the south and apparently alert to his presence, Boest turned north to attack *Arabis*, which he thought had been damaged in his first attack. This second attack on *Arabis* started at 0012 and lasted four minutes. The Germans pumped in gunfire and two torpedoes, this time knocking out *Arabis*' radio and steering with shell hits, but Boest ran east again when he thought the other "cruisers" were coming up from the south and firing on his ships. British records make no mention of the three sloops firing at all, let alone steaming north to take the Germans under fire, so the fire reported was either imaginary or from another German group.[76] Boest took his ships twenty miles eastward, where they met with the two other units of II Flotilla.

A three-boat group led by Schuur in *B97* had reached the farthest extent of its advance at midnight and then turned northeast. Schuur had Boest's deciphered contact report in hand twelve minutes later and saw Mayne's three ships a few minutes after that, just to the northwest steaming southwest. He took them to be Boest's *Arethusa*-class cruisers. The British and German ships exchanged recognitions signals, with the Germans mimicking the British signal. Mayne, unsatisfied with the German reply, turned his ships away. His timing was good, since he saw the flashes of torpedo launches as he turned, and the two torpedoes launched by *B97* at 0022 missed. Confronted with the need to make a hasty attack on rapidly approaching targets, Schuur's other two boats got their firing solutions wrong and either failed to fire or missed.[77]

In keeping with doctrine, Schuur turned east to gather his flotilla and lead a mass attack. On the way he almost engaged friendly boats (probably those of the third II Flotilla group, led by Dithmar) but the German recognition system saved him. The three groups gathered by 0100, whereupon Schuur led them back toward the light buoy. On the way they narrowly avoided another friendly fire incident with a group of VI Flotilla boats. II Flotilla arrived at the light buoy at 0148 to discover *Arabis* still there. The boats closed for the kill, which required still more gunfire and at least another four torpedoes. Boats reported dodging torpedoes and each other as II Flotilla swarmed around *Arabis* and fired from all directions, but two torpedoes hit and finished the sloop. The Germans then came close to yet another friendly fire incident when two groups of IX Flotilla boats steamed to the scene of the action and almost attacked Schuur's formations.[78]

The action showed the awkward realities of night warfare. The German plan laid out a neat operational concept. Eight groups of torpedo boats would advance in a precisely controlled line spread so that no enemy could slip through. Once one group sighted an enemy, the net would enfold it and annihilation would quickly follow. The northernmost groups should have been poised for this, since they were commanded by the officers leading II Flotilla and its two half-flotillas—officers who had served in these roles since 1912 and had a combined forty years of sea duty.[79] The event proved far different. Out of touch with each other, the separate groups could not know their relative positions with precision. Contact with the enemy was reported late, incompletely, and incorrectly. Nor was contact consistently maintained. In this Boest had to answer for the lateness of his report and the omission of an enemy course, but mistakes as to position and the identity of his targets were understandable. Boest's failure to maintain contact was also understandable, given that he thought he faced light cruisers and had lost the element of surprise. Reversing course to attack the fourth possibly damaged "cruiser" by the light buoy seemed the more sensible course, particularly with two-thirds of his torpedoes expended.

Even when in contact with the enemy, commanders struggled to coordinate their forces and weapons. Boest's group fired only three of a possible fifteen shots at Mayne's three ships during its first attack, while Schuur managed only two (or four) out of a possible fifteen. The first attack on *Arabis*, with seven torpedoes fired out of a possible twelve, would have been more credible if any had hit. All credit to Schuur and his commanders and crews for managing a rendezvous of the three II Flotilla groups, but the final mass attack, although textbook in concept, showed the difficulties inherent in close-range night attacks by many ships on the same target. The final assault on the hapless *Arabis* had more the character of mob violence, with *G104* maneuvering radically to avoid torpedoes that, if real, must have come from other II Flotilla boats and with Dithmar's formation sheering off to avoid collisions. Hartog added little to the action, only directing the scattered units to rendezvous for another attempt at the British "cruisers" once the moon had set. *Pillau* never got close to the fighting.[80] Writing in 1912, Kapitänleutnant Rudolf Firle had been dismissive of Japanese torpedo boat attacks in the Russo-Japanese War, arguing that their lack of coordination resulted from inadequate training and the absence of cruiser support.[81] Dogger Bank showed that German training and cruiser support hardly guaranteed a better result.

Torpedo performance was also troubling, with only two hits out of at least twenty-two launches. Boest was unlucky in first firing at Mayne's formation just before it turned away from the light buoy, while *Buttercup*'s sharp-eyed watch officer foiled his second attack, but his group fired seven weapons for no hits against an unalerted *Arabis* standing by the light buoy—the most sitting of ducks. The German official history ascribed the poor performance to the torpedoes being set for deep running. That may account for some misses, given the swell running during the action, but the normal German deep setting of ten feet (three meters) against the sloops' normal draft of eleven feet should have resulted in more hits.[82] Firing ranges were also longer than doctrine prescribed, as theory bowed to the practicalities of combat. Even when firing at longer ranges, maintaining surprise proved chancy, particularly for the big boats of II Flotilla. Beyond that, it was no simple feat to maneuver a formation to a good firing position within a thousand meters of a target, even when the target was a slow-moving sloop. Torpedoes that hit also underperformed expectations. The Germans were incensed that it took two torpedo hits to sink a ship of *Arabis*' size. Although the British prisoners said that *Arabis* only displaced twelve hundred tons, the Germans speculated that her officer complement (reportedly eight officers, when German torpedo boats had only three or four) betokened a much larger vessel. Otto Groos speculated in the official history that perhaps given its intended role the minesweeper had been built with a large reserve buoyancy, which explained why it took "two hits from the newest German torpedo, which carried a particularly powerful explosive" to sink her.[83]

To the good, German night-recognition procedures worked to avert three potential friendly fire incidents, and the boats in the individual groups stayed with their leaders despite some drastic gyrations. Schuur's three scattered elements managed a rendezvous even after two of them had tangled with the enemy. The forethought and training that the Germans had devoted to night torpedo boat warfare had not all been in vain, even if the operational concept proved far harder to execute than anticipated.

BEARDING THE LION

Jutland, fought on 31 May and 1 June 1916, did little to build faith in the nighttime effectiveness of German torpedo boats. Their great moment came in daylight, when boats of III, VI, and IX flotillas launched a torpedo attack that allowed the German battle fleet to escape from a perilous tactical situation. But even then, the attack was a rescue rather than a thrust to achieve a decisive tactical edge.[84] Moreover, it showed that German flotillas still struggled to put the maximum number of torpedoes on a target. And IX Flotilla, despite being in the thick of the daylight fighting, fired no more than thirty-six of the sixty-six torpedoes it carried.[85] As night fell, the torpedo boats were dispatched on various headings to search out and attack the British battleships and battle cruisers, but few contacts resulted. No attacks were made on the big ships, and although the flotillas did expend nineteen torpedoes at other targets from nightfall through the following dawn, they only hit when scuttling cripples: the capsized British destroyer *Turbulent*, the German battle cruiser *Lützow*, light cruiser *Rostock*, and torpedo boat *V4*.[86]

The German torpedo boats did find their foes in a Baltic battle fought on 30 June off Landsort, aided by the bright summer night at high latitudes. But once contact was made, the torpedo boats struggled to get into favorable firing positions, since they began well west and behind the Russian cruiser-destroyer force they were stalking. The five mixed-firing boats in the German force lagged, while the three oil-fired boats just managed to draw even with the head of the Russian force as dawn broke. Seven torpedoes fired during the night failed to connect; another dozen long-range shots after dawn missed as well and the Russians, who until then had been concerned that the German boats were their own destroyers, responded with gunfire. The engagement showed that just finding the enemy did not ensure success. A successful attack required a favorable attack position—something not always easy to achieve.[87]

By October 1916 a renewed emphasis on the war against trade had deprived the Hochseeflotte of its submarine support. With the main fleet hobbled, at least in the minds of the *Admiralstab*, the Germans sent III and IX flotillas to join the boats already operating in Flanders. The Flanders bases offered access to several targets—shipping in the Thames Estuary and the Downs, merchantmen plying the trade route between the Dutch and British ports, troop transports and supply ships in the English Channel, and the antisubmarine mine and net barrier that the British recently had thrown across Dover Strait. Many small fishing boats—drifters—monitored the barrier's nets for submarine activity. They were backed by a few larger vessels that stood by to make radio reports of any submarines spotted. Attacks against any of these targets would force the British to make hard choices about allocating their destroyer forces, which were being stretched ever thinner as the submarine war ramped up.

The two flotillas arrived in Flanders on 24 October. All the boats were oil-burners with six torpedo tubes. The newest had 10.5-cm guns. After long resisting pressure to upgun the boats, the German command had finally decided that larger guns were worth having. The big guns were a substantial upgrade, throwing a 38.5-pound high explosive shell, compared to the 20.9-pound shells fired by the most potent 8.8-cm guns.[88] The first boat built with larger guns entered service in August 1916, while a program to replace the guns in many existing boats, which had begun with the II Flotilla in January, was extended to other boats that summer. Up to half the boats assigned to the latest operation carried the heavier artillery.[89]

Kapitän zur See Andreas Michelsen had taken over command of these two flotillas from Kapitän zur See Johannes Hartog in spring 1916. Michelsen entered the Navy in 1888 as an officer cadet. He gravitated to torpedo warfare in 1897, starting with a tour as a company officer in the Torpedo Inspectorate's training establishment. Although Michelsen never commanded a half-flotilla and only led flotillas in 1906 and 1907, he was heavily involved in training and weapons development in the years following. Notably, from 1912 until the start of the war he served in three roles at once—inspector for the torpedo testing organization, captain of torpedo training ship *Friedrich Carl*, and instructor at the Marine Academy. Michelsen had far less command experience than his predecessor, but far more experience in instruction and weapons development.

Michelsen preceded his flotillas to Flanders, where he worked out a plan for a night raid. Rather than attempting to coordinate many three- and four-boat groups, as in the Dogger Bank sweep, Michelsen planned to operate in four groups of five to seven boats and to keep the groups separated in time and space. The two half-flotillas of IX Flotilla would cruise to the Dover Strait and attempt to disrupt cross-Channel traffic. The five boats of 17th Half-Flotilla with the IX Flotilla leader boat (Korvettenkapitän Herbert Goehle) would keep west of The Ridge shoal and strike at traffic off Folkestone. The five boats of 18th Half-Flotilla (Korvettenkapitän Werner Tilleßen) would stay east of The Ridge and attack traffic off Boulogne. Not only would the two half-flotillas operate in separate areas, they would breach the Dover barrage at different points. After Goehle and Tilleßen had cleared the barrage and reached the strait, III Flotilla, reinforced by three boats from the Flanders destroyer half-flotilla, would raise havoc along the barrier, sinking drifters, supporting vessels, and any transports they could find. These thirteen boats would also operate in two separate groups, with Michelsen taking seven boats from 5th and Flanders half-flotillas across the barrage to the west by Goodwin Sands and the five boats of 6th Half-Flotilla plus the III Flotilla leader boat (Korvettenkapitän Wilhelm Hollmann) breaching the barrier farther east. For both flotillas, a line drawn from Le Sandettie sandbank through The Ridge would demark the operating area of their constituent half-flotillas. Two submarines would be prepositioned north of the barrage to aid navigation, with three more joining later to intercept any pursuing British forces. The torpedo boats were to stop and search any lighted merchantmen on cross-Channel routes and sink any darkened ships on sight. In this time before the declaration of unrestricted submarine warfare, the boats were prohibited from attacking British coastal traffic, which might include neutral shipping.

All did not go according to plan, but the operation was much less confused than the Dogger Bank raid. It began under clear skies in a new moon period with visibility of two to three miles, but with clouds gathering and winds building into a gale by morning. Both groups of IX Flotilla slipped through the barrier unreported late in the evening of 23 October, although Tilleßen's ghosted by a line of four British destroyers steaming from Dover to Dunkirk and then was challenged by old destroyer *Flirt* supporting the drifter line. The lead German boat, *V30*, responded with the usual trick of repeating *Flirt*'s own challenge, which led the Royal Navy destroyer's commander to assume that he was being challenged by the British destroyer group. He obligingly made the proper response and let the Germans go on their way unreported.

The next German units across the barrage were far noisier. Both crossed the barrier at about 2300. Michelsen, to the west, turned farther west along the barrier while Hollmann's group plunged directly into a group of five drifters. The Germans sank three outright in the glare of searchlights and left a fourth

MAP 2.2. DOVER RAID, 26–27 OCTOBER 1916

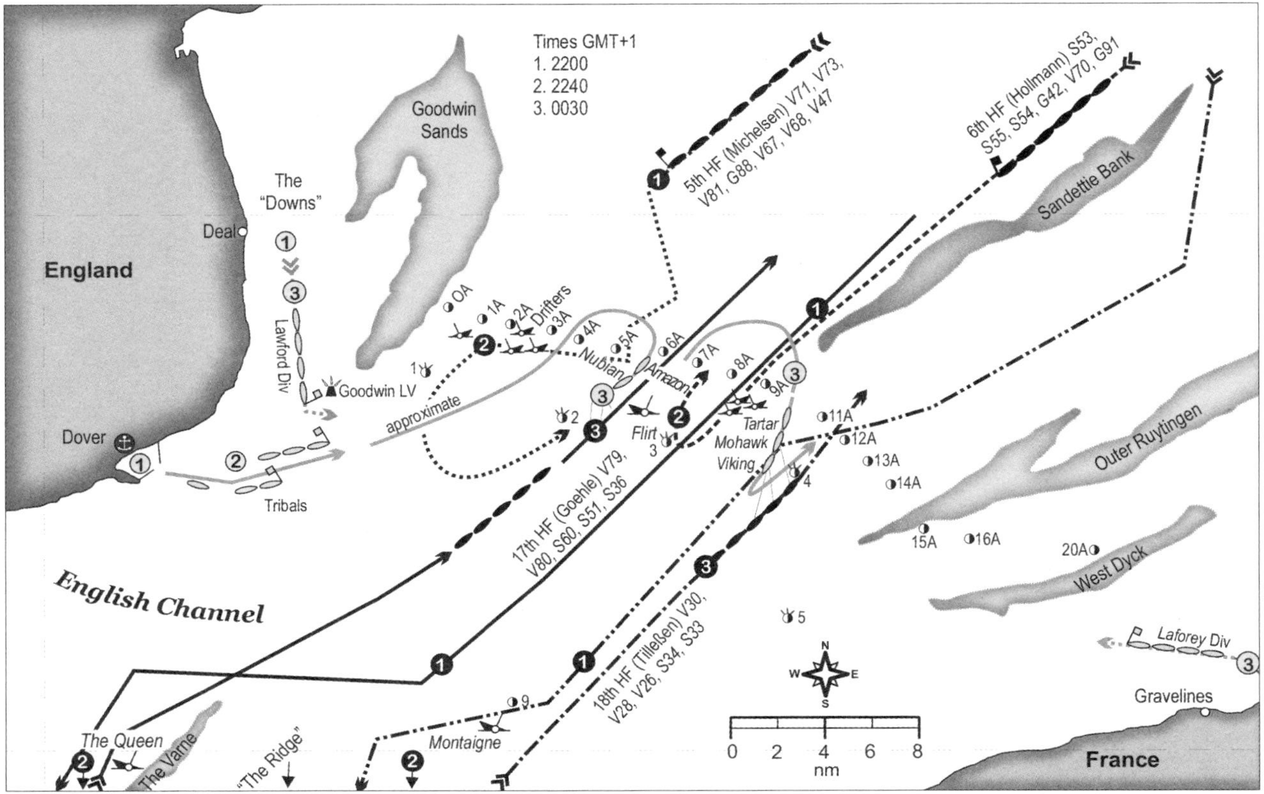

in a murk of steam and smoke from shell hits. *Flirt* heard and saw the action and headed north toward it at 2320. At 2335 she reached the damaged drifter and, with her skipper thinking that a submarine had been the attacker, switched on her searchlight and lowered a boat to rescue survivors. When her crew spotted other ships in the area, her captain thought they were French destroyers. Instead, they were Hollmann's group, which pumped shell after shell into the hapless destroyer from four hundred meters until her boilers exploded. The only survivors were the crewmen who had left her to go on their rescue mission.

Flirt sank without issuing an alert, but armed yacht *Ombra* had also heard Hollmann's guns. She radioed a report of enemy destroyers before trying to shepherd the remaining drifters (none of which had a radio) to safety in Dover. By this time, the IX Flotilla groups were more than twenty miles to the southwest, with Tilleßen probing the waters off Boulogne and Goehle off Folkestone, where he saw the lights of the empty transport *The Queen* just west of The Varne. The Germans stopped the ship, examined her papers, and allowed the crew to take to their boats before *S60* torpedoed her. Goehle thought she sank quickly, but she drifted derelict before foundering off South Goodwin in the rising gale. Tilleßen had less luck, seeing only hospital ships off Boulogne and not even being able to stop them and establish their bona fides due to the gale making up. His group did see two small French patrol craft off Gris Nez, sinking armed trawler *Montaigne* and damaging fisheries vessel *Albatros* before withdrawing. Dover Command got word of the attack on the transport at 0029, giving the first solid evidence that the Germans had penetrated that far into the Channel.

At the barrier, Michelsen was coming into action. Starting at 0015 his big formation engaged two divisions of drifters in quick succession as each tried to run west to Dover. The Germans sank two of the little ships and damaged a third. *Ombra* saw the Germans, but managed to escape without being spotted. Both groups of III Flotilla then turned toward each other to withdraw through the barrier, making mutual sightings but quickly exchanging recognition signals before any friendly fire broke out. Their part in the action was over, but IX Flotilla still had fighting ahead.

Vice Admiral Reginald Bacon, leading the Dover command, had *Ombra*'s alert in hand by 2330. He had five groups of destroyers and a smattering of other ships at his disposal. Four modern destroyers were guarding the merchant anchorage in the Downs and were not available to intercept the

TABLE 2.3. REPRESENTATIVE SHIPS INVOLVED
IN DOVER STRAIT ACTION

SHIP	DISPLACEMENT	SPEED	GUNS	TORPEDO TUBES	PROPULSION
British:					
Flirt	440 tons	30 knots	1 × 3-inch, 5 × 2.2-inch	2 × 18-inch	Piston, coal
Amazon	1200 tons	33 knots	2 × 4-inch/45	2 × 18-inch	Turbine, oil
Laforey	1150 tons	29 knots	3 × 4-inch/45	4 × 21-inch	Turbine, oil
German:					
V26	861 tons	36.3 knots	3 × 8.8-cm/45	6 × 50-cm	Turbine, oil
V67	1169 tons	36.6 knots	3 × 10.5-cm/45	6 × 50-cm	Turbine, oil

Sources: Gröner; Gardiner, ed., *Conway's 1906–921*.

Germans (although they did make a belated run into Dover Strait due to a misunderstanding). Four more modern destroyers—the *Laforey* group—were at Dunkirk. Six Tribal-class destroyers were at Dover and detailed as a rapid response force. Four older destroyers, contemporaries of *Flirt*'s, were sheltering at Dunkirk and would remain there with four old torpedo boats, five monitors, and flotilla leader *Swift*, while four more old destroyers were held in reserve. Light cruiser *Carysfort* was at two-hour notice at Dover. Bacon's challenge was to coordinate these forces effectively.

The Tribals had cleared Dover at 0020, but separated over time into one three-ship formation and three ships proceeding independently. They were drawn to Michelsen firing at the drifters, but did not arrive in time to intervene. *Nubian* saw the Goehle group at 0140 as the Germans approached the barrier but her captain tried to exchange recognition signals before opening fire. The delay was fatal, as the Germans pummeled the ship with shellfire and then blew off her bow with a torpedo. This ended a ramming attempt by *Nubian*, and she drifted off into the rising gale. An encounter between *Amazon* and Goehle's boats had nearly the same result, with *Amazon* flashing recognition signals even as the Germans were firing. Fortunately for *Amazon*, the torpedo launched at her missed, and she reeled off with a gun and two boilers out of action. Tilleßen encountered the three-destroyer group on an opposite course shortly after exchanging recognition signals with Goehle. Again the British challenged, unsure whether the German ships were friend or foe, and again the Germans fired. When the lead Tribal turned to pursue, her captain found his path blocked by the next in line struggling with a jammed rudder.

The British lost contact with the Germans in the scramble to avoid a collision. The *Laforey* group saw the muzzle flashes of Tilleßen firing at the Tribals, but could not make contact with the Germans. Both groups of IX Flotilla steamed home without further incident.

The Germans could count the October raid a success, but not an unqualified success. The raid was undertaken with clear objectives and careful planning. The only damage that the Germans suffered resulted from a collision between *G91* and a damaged drifter. They sank outright or caused the loss of a transport, seven drifters, two destroyers, and a trawler. They also damaged another two destroyers, two drifters, a trawler, and a patrol vessel. In all, more than half the vessels supporting the barrage had been sunk or damaged. Recognition procedures had proven effective in the two close approaches of separate German formations and they had effectively exploited the weaknesses of the British recognition systems. The patrolling British ships were much more accustomed to seeing unanticipated friends than the enemy, which explains *Flirt*'s failings, but even the Tribals, alerted to the presence of the enemy but unsure of the location of friends, signaled rather than shooting. Goehle and Tilleßen had passed the barrage without triggering an alarm and had penetrated as far as Folkestone and Boulogne. But the raid was not flawless in concept or execution. Although the Germans avoided friendly fire incidents, the two close approaches of their own forces were not planned. While the attack on the barrage distracted attention from the raid on the shipping lanes, it also resulted in IX Flotilla facing alerted British forces when its boats recrossed the barrage. The free use of gunpower enabled the German boats to make short work of smaller craft and *Flirt*, and to inflict significant damage on other vessels, but the fleeting nature of night actions made it difficult to sink the Tribals. Keeping the IX Flotilla boats concentrated in two groups to facilitate control limited their ability to search for merchantmen. IX Flotilla only spotted three of at least sixteen merchantmen crossing the Channel in its area and two of those were brightly lit hospital ships. Even *The Queen* had her navigation lights burning. Another forty ships were plying the waters farther east, beyond the scope of the raid. Loaded transports, the highest priority target for the raiders, had already been banned from sailing at night. In all, the German planning and execution of the Dover raid showed significant advances over previous attempts but also underscored the frustrations of achieving significant results in night actions. The most consequential result of the raid was not in the damage that it inflicted but in the diversion of resources that it provoked. The British detailed more destroyers to reinforce the Dover Barrage—ships that would be sorely needed when the Germans resumed unrestricted submarine warfare in February 1917.

The October raid began an active period for German torpedo boats. In the following two years they mounted two dozen night operations from Flanders and another dozen from North Sea bases. Some of the North Sea operations were like the Dogger Bank sweep, although the targets now were mainly merchantmen rather than light patrol forces. These sweeps saw little success. The Belgium-based operations, as well as half of those mounted from the North Sea, focused on five target areas—the British coast around Lowestoft, the Downs, the sea lanes from the Netherlands to the British ports, the Channel, and the coast around Dunkirk. Overall, the Germans took a steady toll of British and French patrols and occasionally sank or captured a merchantman, but none of the raids produced or could have produced decisive results. Forays were risky if prolonged, as in April 1917, when six boats lingered and lost two of their number to British forces. Merchant traffic was hard to find, due to stringent Allied traffic controls and a lack of reliable intelligence. A German strike could occasionally savage the small craft outposting the barrage line, but these were quickly replaced. Although Vice Admiral Bacon professed to fear a concentrated German operation against the many merchantmen sheltering in the Downs every night—and the Germans did send torpedo boats to the Downs on six separate occasions—the boats never pushed past the entrances and generally contented themselves with desultory bombardments of the shoreline.[90] They were barred from attacking British coastal traffic until unrestricted submarine warfare began. Then, as the British stiffened the defenses of the Downs with monitors and shore batteries, any opportunity to strike a substantial blow disappeared.[91] In fact, the chances of striking any blow at all began to shrink as the war entered its fourth year. The Allies stepped up their sea and air bombardments of the Belgian ports and their in-depth defense of Dover Strait with constant illumination, and the Germans found all their energies absorbed in combating intensive Allied mining and barrier campaigns. The last raid in the Dover Strait took place on 15–16 February 1918 and the last raid on the Dunkirk coast—conducted by motor torpedo boats, with the larger torpedo boats providing security—on 22–23 August. Neither strike had any success.

Tactical concepts die hard, and the final planned operation of the Hochseeflotte put torpedo boats into two of the roles that had been envisioned for them before the war. The British fleet would be drawn into action by torpedo boat and light cruiser sweeps, with II Flotilla to operate off the Dutch coast and 2nd Half-Flotilla to accompany light cruisers to the Thames. And if that failed to provoke a reaction, the torpedo boats were to sweep directly toward the British base in the Firth of Forth, where, in conjunction with minefields and submarines, they would deplete the British before the climactic battle.[92] As it turned out, mutiny flashed through the fleet before the Kaiser's ships could steam to their last desperate struggle.

CONCLUSION

The German navy was not alone in struggling to fight effectively at night. The sinking of Austro-Hungarian dreadnought *Szent István* by tiny Italian motor torpedo boat *MAS-15* was the outstanding success in World War I night torpedo attacks. No other navy sank a dreadnought in this way. Night attacks by German torpedo boats sank a monitor (later salvaged and returned to service), three destroyers (one salvaged), a minesweeper, a minelayer, and two

PHOTO 2.4. A daylight role. A *G37*-class destroyer with a bone in her teeth darts between two battleships. Seeking a role for torpedo boats in daylight battle, the Germans practiced maneuvers in which flotillas would cross through the battleline to deliver torpedo attacks. Tactical thought focused on this new mission, even though night attacks remained the primary task of the torpedo force. *U.S. Naval History and Heritage Command, NH 45402*

merchantmen—a paltry toll compared to sinkings by German submarines, but broadly consistent with results in other navies. The greatest successes came in two classic roles—an action against a blockading force, in which *S90* sank *Takachiho*; and a harbor raid on 19 October 1917 during which four small coastal boats put three torpedoes into monitor *Terror* as she rode at anchor off Dunkirk. But there were few blockaders to attack, and harbor-raiding was a perilous pastime. Perversely, the prewar increase in the numbers of German torpedo boats and in the effectiveness of their weapons, along with other factors, prompted the Royal Navy to adopt a distant blockade that kept targets out of the German navy's reach.[93]

The German torpedo boats lacked tools that would prove crucial for night fighting—voice radio for command and control, radar and effective night optics to spot and track both enemies and friends, and procedures for plotting ship positions to maintain a clear tactical picture. Of these tools, radar was for the future (although a primitive radar device was offered to the German navy in 1916), while voice radio was first introduced by the U.S. Navy at the end of the war. The British navy pioneered plotting during the war, but not for destroyers. No navy fully recognized the potential for improved night optics. German industry had deep resources in both radio and optics technology, and likely could have produced voice radio and superior night optics had they been demanded, but, with so much of their attention focused on day battle between capital ships, naval leaders did not identify these needs.

The German torpedo boat force began the war with tactical and operational concepts necessarily shaped only by theory and peacetime exercise. The navy saw night operations as essential to its strategy, but night-fighting doctrine stalled as the battleship fleet grew and torpedo boats took up a role in day battle. Although operational concepts called for torpedo attacks on light blockading forces, tactics focused on mass attacks against enemy battlelines. War revealed flaws. Opportunities for night torpedo attacks on enemy battlelines were exceedingly rare, with skirmishes with opposing light forces far more common. The Kaiserliche Marine struggled to find enemies at night or shadow them at dusk. Coordinating attacks from widely spread scouting lines proved difficult. The failure to find *Slava* foreshadowed Jutland, where the Hochseeflotte lost contact at dusk and its torpedo forces could not regain it despite searches down various compass bearings.

Even when contact was made, night-attack doctrine proved largely unworkable. Insistence on attack ranges of six hundred meters or less failed to recognize the need to get torpedoes onto a target before the target knew they were coming. Successes at short range almost always required a distracted enemy or one uncertain of whether the attackers were friend or foe. The tactical manual was not wrong in recognizing the difficulties inherent in torpedo fire at night, but its response was flawed. Further, the directive that no torpedo be wasted clashed with the need to blanket an opponent with enough torpedoes to make at least some hits. Prodigious use of an expensive weapon was unpalatable but necessary. The navy continued to explore daytime long-range torpedo fire during the war, concluding that fire at individual targets beyond three thousand meters was futile (a conclusion that the British had reached before the war) and experimenting with long-range coordinated fire against enemy formations.[94] It also developed techniques for a submarine to fire a salvo of up to four torpedoes, perhaps spurred by the accumulated experience of more than nine thousand submarine attacks carried out in the first two years of the war.[95] But surface torpedo fire at night largely remained a matter of a boat shooting one or two torpedoes at a time. Although experience showed night-attack doctrine to be faulty, the physical layout of the torpedo armament, the lingering effects of the "don't miss" and "save the torpedoes for the big ships" doctrines, and the idea that the cumulating point of an attack would be a pell-mell dash at the enemy all weighed against the development of massed surprise attacks as a tactic in night combat. Worse, results were lacking even when torpedoes were fired at "can't miss" targets—witness the seven fired at *Arabis*. All this saw German torpedo boats shift from torpedoes to guns as the war went on.

Experience revealed problems, but not ready solutions. The problem of coordination between formations was dealt with by dispensing with the need for it. The raids against the British and French coasts relied on formations of three to seven boats well separated to give each group the ability to immediately engage anything it saw. Separation was not always perfect, but it was good enough that, when combined with the excellent German night-recognition system, friendly fire incidents were avoided and the Germans generally got in the first shots against an enemy. German crews proved to be good at maintaining formation at night and in adverse conditions. German torpedo boats and destroyers made good use of their increased gunpower. But small formations, combined with fleeting encounters and short visibility ranges, meant that German torpedo boats were seldom able to annihilate an enemy. While upgunned boats were able to inflict more damage, and torpedoes fired from short range occasionally hit home, the scope for torpedo action remained limited. Changes in tactics and in armament meant that German torpedo boats became a more potent force as the war went on, and one that often surprised and bested its opponents, but it never achieved the results that had been envisioned for it before the war. For the Germans, throughout the war torpedo boat tactics would be the tactics of frustration.

CHAPTER 3

THE BRITISH AND NIGHT FIGHTING AT AND OVER THE SEA, 1916–1939[1]

JAMES GOLDRICK

The successes of the Royal Navy in night fighting during the 1939–1945 war stand in stark contrast to the deficiencies of its performance between 1914 and 1918—a difference they usually attribute largely to the development of radar. Undoubtedly, radar changed night fighting profoundly. But the achievements of World War II were built on a much wider basis of technological development, combined with tactical innovation, and deliberate efforts to cultivate a culture of the tactical offensive. This chapter traces the course of British night-fighting doctrine and activity from the aftermath of the Battle of Jutland until the beginning of World War II. It concludes before the advent of effective surface search radar, a development that occurred only after 1939.

The dead hand of economy lay heavy on the Royal Navy during this period, but the service had certain advantages. The first was time, at least between the mid-1920s and mid-1930s. Technology also developed at a much slower pace than the breakneck speed of the decade before World War I. As a result, there were opportunities to innovate and experiment in a more controlled way than had been possible before 1914 and to build steadily on what was learned. The second was the wealth of hard-won operational experience that permeated the middle and senior ranks during the entire period. A thirty-two-year-old lieutenant commander in 1914 could be a forty-two-year-old captain in 1924 and a fifty-two-year-old rear admiral in 1934. A seventeen-year-old midshipman at sea in 1914 would be only thirty-seven in 1934, perhaps by then a junior commander still to be considered for promotion to captain.

A JUTLAND VIGNETTE

The Royal Navy's experience of night fighting in the World War I fell into two distinct categories: main fleet engagements, and the encounters—usually confined to light forces—associated with the attack on and defense of coastal shipping or the English Channel. Nevertheless, the systemic failures with which Jutland is usually associated had their counterparts elsewhere in the war of the small ships. Here is an extract of notes written by Lieutenant Commander E. R. Corson, navigator of the light cruiser *Caroline*, on 1 June 1916:

> The 4th LCS, with ourselves as tail ship still, was ahead of the Battle-Fleet. And so the night passed away. Everything was dark—no lights of course.
>
> But there was just enough luminosity to see and sometime in the Middle Watch the ghostly shape of a battleship loomed up a quarter of a mile or less away on our port side, again showing no lights. The Captain looked at her, apparently calmly. I, looking too, had to be calm as well but inside I was terrified. We were supposed to be five miles or so from any battleship. If she was British would she take us for a German and open fire? The seconds dragged away. Silence . . . silence! She drifted by . . . faded away. I do not suppose we shall ever know who she was but those few minutes were the most nerve-wracking of the whole affair.[2]

Corson's plaint covers many of the problems that the Grand Fleet encountered during the night fighting of Jutland, and reflects much of the total experience of the Great War at sea. The list includes uncertainty about the disposition of friendly forces that contributed to *Caroline*'s failure to identify the lone German battle cruiser *Moltke*; the fear that friendly fire could be not only delivered but received; and the lack of any impetus to report the sighting, even in a unit whose primary role was scouting. It is also significant that the incident does not seem to have been recorded in *Caroline*'s fair log—or brought to the attention of other officers in the ship.[3]

MAIN FLEET ENGAGEMENTS: JUTLAND'S POST-MORTEM

The first lesson taken from Jutland would remain the focus of main fleet attention for more than a decade: "the training of the light forces . . . had not brought out the need for close and rapid organisation of the light forces towards the end of daylight. Thus, sufficient information was neither given to nor received from them to produce efficient night attacks. They did not keep touch with, nor search for the enemy battle-fleet, and such attacks as developed with one exception were accidental rather than intentional."[4] What went largely unsaid, but seems to have been obvious at the time, was that the essentially defensive role imposed upon the destroyer flotillas by Admiral John Jellicoe, in the *Grand Fleet Battle Orders* (GFBOs), contrary to the approach of his predecessor, Admiral George Callaghan, as commander in chief of the Home Fleets, had contributed to the collective failure to seize all the opportunities offered during the night. It had also meant the absence of enough reports to give a tired fleet commander sufficient indication of just what was happening astern of his battle squadrons.[5]

Existing *Grand Fleet Battle Orders* had been explicit in directing destroyers to attack the enemy with torpedoes at night *only* if the main action had already been decisive. Otherwise, the priority was to prepare the battle fleet for a renewed action at dawn. Moreover, any attack on the enemy, day or night, was to be left to the First and Second Light Cruiser squadrons and the Harwich Force, which were to attack the enemy's light forces as well as its larger vessels. All that said, however, the light forces' *highest* priority still was to maintain contact with the enemy main body.[6]

After Jutland, the Grand Fleet issued doctrinal reforms that required commanders to assemble an "Attacking Force" of destroyers and light cruisers at dusk following any battle fleet encounter with a primary mission of assaulting large enemy vessels during the night. The changed guidance also provided more detailed direction than they had in the past for the command and control of these forces and for the units' reporting responsibilities. The changes also included more emphasis on the need for other scouting units to remain in contact with the enemy and to provide the reports needed to give the Attacking Force sufficient situational awareness to fulfil its new mission while the main body prepared to resume action at dawn. In truth, however, the revised GFBO—known as No. XIII: Orders for the Conduct of the Fleet after Action—was too detailed and complex. It attempted to provide for practically every situation.[7]

Later revisions in 1917 made few modifications, despite the efforts of the new commander in chief, Admiral David Beatty, to decentralize and simplify. As Beatty himself assessed just after Jutland, leaving the destroyers to their own devices was "doubtful of attainment and it is possible that our destroyers searching for the enemy may meet some of our light cruisers trying to keep touch with him."[8] This weakness would be demonstrated repeatedly during postwar exercises. Even with much smaller destroyer forces and fewer cruisers, organizing the transition from supporting the battle fleet during a day action to reconfiguring for attack at night was complicated and hard.

By 1917 all units above destroyer size in the Grand Fleet had been instructed to maintain their own plot of enemy contact reports and other relevant information. But implementation was not immediate. Captain H. R. Crooke, who remained in command of the light cruiser *Caroline* until the end of March that year, claimed when captain of the battleship *Emperor of India* in 1921, not only that, "[t]here is no information at present available in this subject [tactical plotting] in *Emperor of India*," but also [an annotation made in his own hand] that "[t]he ships in which I myself served during the War had no plotting arrangements."[9]

There were other surprises at Jutland. One was that the German units that switched on their searchlights and engaged attacking destroyers with both secondary and main armament created many more difficulties than those that remained darkened and turned away. Later analysis noted that the destroyer captains greatly preferred the latter system "as they were able to attack with deliberation and were hit very little." By contrast, being engaged by 28-cm as well as 15-cm guns was found "very trying to the destroyers." An officer of the destroyer leader *Broke* commented of his ship's engagement with the German battleship *Westfalen:* "It was perfectly damnable having their light right in our faces (about four or five cables off) and being properly biffed."[10] The British also realized that not only was the German star shell effective in illuminating an attacking force, but it did not reveal the position of the units that fired it. This spurred the Royal Navy into the immediate development of star shells. Although the initial effort was concentrated on munitions that could be fired by secondary surface armaments, gunnery specialists quickly realized that the proliferating antiaircraft (AA) weapons provided an effective alternative—especially since they had no other practical utility at night. An associated innovation that made star shells even more attractive for night fighting was flashless powder, which reduced the chances that flashes from the weapons would reveal the British forces to the enemy. Such technological improvements helped change attitudes. By 1926, star shells were of sufficient value that the modernized battleship *Queen Elizabeth* included a hundred rounds for her 4-inch guns and a hundred more for her 6-inch secondary battery. The fact that 7 percent of the total 6-inch shells carried were star shells clearly indicated the importance the Admiralty accorded to star shells.[11]

Notably, however, flashless propellants were not fully extended to larger-caliber weapons in later years. At the Battle of Surigao Strait in October 1944, the Australian heavy cruiser *Shropshire,* firing British pattern shells and propellant, found herself the only ship not firing flashless ammunition in the otherwise all-American battle line.[12] This was because the flashless propellant initially developed for the Royal Navy was too bulky for guns larger than 6-inch (and not fully "flashless" at that caliber), which made the use of the lighter AA weapons for star shells even more attractive.[13] The other factor militating against the extension of flashless propellant was a genuine belief within the gunnery division of the Naval Staff that the psychological impact of unexpected muzzle flashes had real value in further unsettling an already-surprised enemy.[14]

Searchlights also received attention. Control arrangements were substantially improved to enable ships' crews to direct them from a single position. Another innovation was the development of searchlights that could be kept on with shutters closed and emitting no light, enabling crews to open them quickly to illuminate a designated target.[15] Centralized remote control permitted more effective employment of searchlights. And the new installations were deliberately sited on after-superstructures when practicable, to keep the enemy's point-of-aim away from the bridge and gunnery-control positions and to minimize the dazzle effect on the personnel there.

What eventually became clear was that star shells and searchlights were complementary systems rather than alternatives. As one gunnery expert wrote during World War II, based on his interwar experience: "In practice, therefore, it is usual to try to reap the advantages of both, that is, to begin firing with searchlights so as to obtain the *instantaneous* production of maximum output, firing star shells at the same time, and, when the star shells are seen to be effective, switching off the searchlights, which can always be resorted to if needed."[16] Doctrine also provided for using searchlights when facing a close-range attack by enemy destroyers, "where the dazzling effect of a searchlight should be exploited to the full."[17]

The Admiralty's first attempt to improve British recognition procedures was to issue cruisers red, blue, and green shades for their signal lamps "to

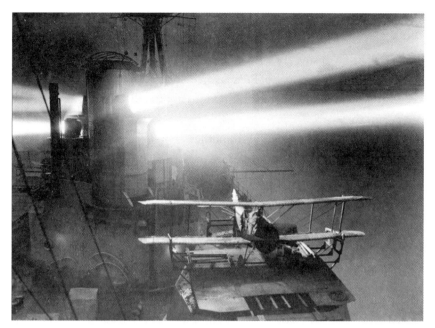

PHOTO 3.1. HMAS *Australia*, demonstrating the blinding brilliance of her searchlights look-ing aft from the bridge. This photograph was taken in 1918 when she was serving as flagship of the Second Battle Cruiser Squadron in the Grand Fleet. Note the Sopwith Strutter aircraft on top of P turret in the foreground and Q turret to the back left. *Sea Power Centre Australia, via Dr. David Stevens*

make the Private Signal Challenge and Reply more distinctive and less eas-ily copied by the enemy."[18] However, although the British tried to find more efficient challenge-and-reply systems, the more important realization was that "It is considered essential that all ships should extemporise arrangements for training their searchlights as well as their guns dead on to a suspicious vessel, ready for instant attack, before making the challenge."[19]

LIGHT FORCE ENGAGEMENTS:
NIGHT FIGHTING IN THE NARROW SEAS

The experience of night fighting by the Dover Command and the Harwich Force, largely defensive, provided repeated examples of the disadvantages inherent in that role. Maintaining the necessary alertness night after night in overworked patrol and escort units was practically impossible, even in periods of known threat. The difficult navigational conditions in the English Channel, with its strong currents and frequent bad weather, also meant that units could often not be certain that other, friendly ships were in their assigned stations, even when they had been given full information about local movements, which was by no means always the case. The situation had been complicated even further by the presence of French forces in the eastern areas of the Channel—with each of the Allied navies anxious not to mistake the ships of their partner for Germans.

The Dover Barrage itself had another complication for its defenders. Since the primary targets of the barrage were the U-boats attempting to pass through the English Channel, a surface night attack using torpedoes could be mistaken for a submarine attack. On 17 March 1917, British destroyers went to the rescue of the first victim of a German surface sortie into the Channel. Revealing their presence by the use of searchlights in their hunt for survivors of what they thought was a U-boat attack, they became new prey for the German torpedo boats.[20]

Poor standards of training had been another bugbear, a situation not helped by the British practice of dispatching destroyers and other light craft on operations with little or no preparation after commissioning. Time and again, onboard drills failed or serious material defects emerged under the pressure of unexpected combat. Conditions on the newly refitted minesweeper *Newbury*, attacked by German torpedo boats in the English Channel early on 15 February 1918, were so disorganized that sailors could not readily find the green flares that were meant to warn other units of the enemy's presence. As a result, the Germans were able to sink several other craft and escape.[21] The need to provide dedicated training for small ships so they could imbue their inexperienced crews with the basics of their job was a lesson that the Royal Navy was to take into the next war.[22]

What became clear from battle experience was that immediate reaction to a nocturnal sighting provided a decisive advantage, giving the potential of overwhelming the opposition before it could fire a shot. It was especially clear that "when vessels are isolated at night and resort to challenging" a perilous situation was created. A succession of German successes against the Dover Patrol put the practice of issuing the challenge further into disrepute. In March 1917 the Flag Officer Dover ordered patrolling units to open fire immediately if the balance of probabilities suggested that the contact was hostile.[23] Existing Grand Fleet doctrine already instructed that a challenge could not be issued until the unit concerned was ready to open fire at the first indication that a contact was hostile—a practice adopted for the entire navy after the war. Tragically, the difficulties that the Royal Navy of 1914–18 encountered in ensuring that doctrinal changes were not only promulgated but understood were starkly demonstrated during a convoy attack by German destroyers in the Norwegian Sea on 12 December 1917. The destroyer *Partridge* was overwhelmed, in part because of time she wasted making the challenge.[24]

The number of units that could work together without confusion depended upon several circumstances, most notably the visibility in a region in which fog was as much of a problem as darkness. There was evidence, however—most clearly shown by the action of 21 April 1917—that a pair of fast and heavily armed units accustomed to operating in company provided an excellent combination of strength and agility. The destroyer leaders *Swift* and *Broke* took heavy punishment from the German torpedo boats that they faced, but the combination of torpedo and ram they employed destroyed two of the enemy force.[25]

Another facet of night fighting in the English Channel arose with the introduction of coastal motor boats (CMBs). These fragile craft were extremely constrained by weather and proved highly vulnerable to air attack as well as to surface gunnery in good visibility. Their limitations meant that they were not only more suited to night operations than to daytime work, but also to harbor attack rather than engagements in open waters. This proved the case on the night of 7 April 1917, when a group of CMBs attacked a force of German torpedo boats at anchor off Ostend. Their inevitably noisy high-speed approach was partly disguised by a seaplane raid on the port, and the CMBs succeeded in sinking *G88*; another torpedo that hit *V81* failed to detonate. The Germans initially thought they were the victims of a submarine attack, and later decided that the CMBs owed their victory largely to having been totally unexpected.[26] Significantly, during a night raid later the same month that resulted in the destruction of the French destroyer *Étendard,* the German torpedo boats were convinced they had been attacked by CMBs in the approaches to Dunkirk, although no such vessels were at sea at the time. The CMB force, equipped with larger and much improved boats, would go on to have a series of further successes against the Soviet Navy in the Baltic, sinking or disabling a number of major units during night raids on the Soviets' major base at Kronstadt.

POSTWAR NIGHT FIGHTING DEVELOPMENTS

The disruptions of the immediate postwar period, notably the operations in the Baltic and in the eastern Mediterranean and Black Sea, as well as the demands of demobilization meant that there was relatively little progress in most areas of tactical development in the early 1920s. When circumstances permitted the resumption of tactical exercises and experimentation, the focus continued to be on main fleet encounters, but with the explicit intent to exploit the night to a much greater degree than ever before. The judgment remained that "Night actions between heavy ships are usually not desirable," but the potential for light forces to prevent a retreating enemy from reorganizing was obvious. So was the desire to continue improving British capabilities.[27]

By the mid-1920s, the groups of destroyers and cruisers assigned to attack or search for the enemy fleet at night had been redesignated the Striking Force and the Searching Force respectively. Their distinct roles were more clearly defined (although the term "Attacking Force" often reappeared in reports and analyses in later years). Exercise A.C. in 1930 confirmed the problems that the commander of the Striking Force faced in trying to control ships once they had encountered the enemy. The night experience also supported the evidence of daylight exercises, which demonstrated that the three-ship destroyer division was a particularly flexible formation in "difficult conditions."[28]

A new area for development came with the use of aircraft for torpedo attacks. The first use of flares to illuminate targets came as early as 1920 and the first formal night torpedo attack trials were conducted in 1923. These early experiments confirmed the potential of such operations by night in certain circumstances, and identified the equipment modifications that would be necessary to allow safe flying over the sea at low level in darkness.[29] However, further development would have to wait until the Royal Navy could deploy sufficient aircraft in the carriers that were being built or awaiting conversion; the Royal Air Force (RAF) had too few planes in the handful of shore-based units in its Coastal Area.

Shadowing

Night shadowing received much attention. By 1927, confirmed in Exercise K.F. in March that year, it was apparent that intelligent exploitation of the prevailing conditions—notably the state and position of the moon—could help a shadower cloak his own presence while maintaining contact with the enemy.[30] It rapidly became clear, however, particularly in Mediterranean conditions, that large cruisers were much more easily detected at night than smaller units were. Although this discovery was made with heavy cruisers of the *Hawkins* class, there were concerns that the high-sided, ten-thousand-ton *Kent* class, built to the limits allowed by the 1922 Washington Naval Treaty, would be even more vulnerable, which probably contributed to contemporary criticism about the new type. Within a few years after the *Kent*-class ships entered service those judgments were confirmed. The *Kent* and *London* classes were "very large and easily detected at night" and had limited utility in most night-fighting scenarios.[31] In the event, despite the desirability of having 8-inch gun units to match the Italians (and the importance their commander in chief attached to them during the Abyssinian crisis of 1935–36, in which he described them as "the decisive factor"), few of the big cruisers spent much time in the Mediterranean during the war.[32] Instead, they would operate in theaters such as the Norwegian Sea and the Indian and Pacific oceans, for which their weatherly qualities and endurance made them much better suited.

The increased use of night shadowing brought inevitable countermeasures. Although the preferred position for a shadower was astern of the force being shadowed, that advantage had to be weighed against the increased risk of attack. Shadowers were thus enjoined to change their position and make their reports at irregular intervals. Since the technology did not yet exist to allow onboard high-frequency (HF) direction-finding (DF) at this point in the 1920s, it was better, according to formal doctrine, for a shadower to use HF rather than the much more readily "DF-ed" medium-frequency band for its reports. On the other hand, making smoke to confuse a shadower proved to be an effective mechanism "if continued with moderation for some time. The result is perplexing to shadowing vessels who are uncertain of the course in the gathering gloom."[33]

In 1930, Exercise O.X. successfully employed the technique of using a scouting unit to illuminate the target formation by searchlight in order to provide aiming points for the approaching battle fleet. Since this meant that the attacking heavy units needed to do nothing that might reveal their presence before opening fire, it was an attractive tactic and would be employed repeatedly in later exercises. The risk was that, as with any use of searchlights, the illuminating unit revealed its own presence and was in danger of being immediately engaged and disabled. This meant that the illumination should only be initiated when the shadower could be sure that its battle fleet was ready to shoot. During Exercise O.X., although the illuminating units were so attacked, their illumination was assessed as being available long enough for their battle fleet to engage at 16,000 yards in what in any case must have been exceptionally favourable conditions of visibility. During Exercise R.O., in 1933, however, "In the only instance of its use the illuminating ship was promptly engaged and sunk."[34] Admiral Dudley Pound, commander in chief of the Mediterranean Fleet in 1936, commented regarding one encounter exercise, "If one makes a mistake one's cruisers get blown out of the water for nothing. . . . The Destroyers found the Battle Fleet by directional bearings of the shadowers but neither this nor the plot is I think sufficiently accurate for a battle fleet engaging the enemy battle fleet."[35] Night fighting remained an uncertain art.

Direction-Finding on Radio Intercepts

It became increasingly clear that using direction-finding gear to track intercepted radio transmissions provided an important addition to a force's ability to develop situational awareness at night as well as in daytime. Both the fit and the capability of DF systems in British ships increased progressively during the interwar years. Continuing research by a small team at the Royal Naval Signal School fostered improvements in detection probability and accuracy and extended the range of detection into ever-higher frequencies—an area in which Britain was well ahead of other countries and took some trouble to conceal its success.[36]

The obvious benefit of having multiple DF sets in a searching force was that they could triangulate enemy transmissions and provide a rapid situation report with the position of the units involved and with a reasonable degree of accuracy. What became evident—and was particularly valuable for night operations—was that DF also could be used to confirm the accuracy of the reported position and thus evaluate the enemy reports of one's own scouting units. Given that the scouting units were required to break radio silence to make a contact report, this made a virtue out of necessity. There were plans to extend the practice to reconnaissance aircraft, which had problems in the 1920s reporting their dead-reckoning positions accurately enough for tactical decision-making (as they had had in 1914–18), but by 1935 aerial navigation techniques had improved to the point that this was not considered necessary.

Direction-finding suggested one anti-shadowing measure: understandably, a shadowing unit tended to remain on a steady relative bearing in the stern arcs of the force being shadowed. Its transmission of regular contact reports created the possibility that accurate DF bearings, combined with the associated estimate of course and speed to provide the shadower's likely future position, could be used to dispatch torpedoes at the target. The restrictions of peacetime meant that the tactic was rarely tested with running weapons, but such an experiment occurred at least once.[37] In practical terms, the shadower was more at risk of approaching too close to the force being shadowed, either through underestimating the range or being slow to appreciate that the target's course or speed had changed—sometimes with the deliberate intent to create such a situation.

Recognition

Recognition procedures remained a problem that defied complete solution. The Admiralty established a committee to examine the issue again in 1926.[38] The Naval Staff became interested in the possibility of a "scientific" solution, most notably through the development of coded infrared signals, which could be transmitted without alerting an enemy. The truth was, however, that the delay in engaging inherent in any attempt to make a challenge and receive a response usually gave the enemy the advantage of being the first to shoot.

Situational Awareness through Near-Real-Time Plotting

The potential for achieving the vital understanding of the location of both friendly forces and the enemy was considerably increased by the development of improved plotting techniques, particularly the plotting "tables" that had begun to emerge during World War I. The transmission of gyroscopic compass and log information to a display that could show the movements of the ship concerned in near-real-time was vital in creating a coherent picture from external reports. In 1924 the Admiralty Research Laboratory (ARL) had begun developing a more accurate navigational course plotter that would be the core of a new plotting table. After trials in the battle cruiser *Hood* and battleship *Warspite*,[39] the complete plotting table went to sea in the fleet flagships *Queen Elizabeth* and *Nelson* in 1929, at the same time that a rival version, the Brewerton Mark 1-A plotting table, was being issued to thirteen major units in the Mediterranean and Atlantic.[40] By 1933 the ARL table had proven its superiority. It would be standard equipment in the Royal Navy for many years. Ironically, one of the factors in making the ARL table the preferred system was its speed clock, which provided a supplement to direct log input, the accuracy of which could not always be relied upon (and on which the ARL did separate work in the same period).

The ARL table had another important advantage—something as apparently basic as "placing [the] plotting paper above the recording instrument itself."[41] The position of the moving ship was continuously displayed as a tiny circle of light, which, projected upward, was visible through tracing paper placed on the table's glass top. This and the track record that could be produced made the job of the plotting officer easier and faster, with obvious benefits for those who were trying to make sense of the resulting picture. Plotting systems also were fitted in cruisers, initially the earlier Brewerton Course Recorder, since it was essential that scouting units in the Searching Force had good situational awareness.[42] Simpler systems were provided to destroyer leaders and destroyers.[43]

The provision of better systems for maintaining the plot was coupled with improved procedures for reducing any *relative* errors between reporting units and flagships, for these were the errors that mattered. Such systems, developed during the 1914–18 war, centered on the promulgation of a reference position by the flagship at regular intervals, on which all succeeding contact reports would be based. Wartime needs had led to improvised manning, as well as physical arrangements to support and rationalize the operation of the plot. The Admiralty conducted a survey of major units in 1920 to understand what had been "best practice."[44] Notably, this only occurred because of a prod from Captain Reginald Plunkett-Ernle-Erle-Drax, then director of the Staff College, who feared that key lessons and techniques learned during the war would otherwise be lost. Drax, a veteran of Beatty's wartime staff—who would be one of the driving forces in the Royal Navy's interwar tactical renaissance— was quite right, as indicated by several of the responses from major units, such as that of Crooke cited earlier. Although funds were short, a formal effort was made in 1923 to standardize the arrangements in battleships and cruisers based on assessed best practice, particularly in the allocation of manpower.[45]

There were some simpler approaches to improving awareness, which included attention to lookouts. Stationing personnel on the upper deck, well below the usual lookout positions around the bridge and on masts, sometimes provided earlier detections at night, since ship silhouettes tended to show up more clearly when viewed from closer to sea level. Providing the best types of binoculars was another. Barr & Stroud's CF41 binoculars, specially developed for night work, became standard issue in the Royal Navy in 1935.[46] Rear Admiral Max Horton, a distinguished submariner commanding the First Cruiser Squadron in 1935, emphasized the importance of alert lookouts to the point that the Mediterranean Fleet destroyers found it very difficult to approach his heavy cruisers—despite their size and profile—without being detected. In speaking to the crew of *Shropshire*,[47] Horton stressed, "At night, the one who sights first is three parts of the way to winning the battle."[48] Modern instructional techniques were also enlisted. In 1937 the Admiralty issued a silent film on controlling night torpedo firings based on Exercise S.N. (a night attack on an unscreened battle fleet).

MOVING TO A NEW APPROACH TO NIGHT FIGHTING

By the late 1920s a formation commander could be much more confident that he knew where his own ships were and thus whether any new detection was of a hostile force. It was this improved situational awareness that was at the heart of the changes that were coming in night-fighting doctrine, and which would provide the opportunity for Britain's new generation of fleet commanders to make them happen.

There were other factors at work, however. One was the growing understanding of and confidence in night warfare generated among the seagoing forces as the 1920s went on, aided by environmental conditions that were usually much more favorable than had prevailed in the North Sea theater during World War I. This was particularly true in the Mediterranean. A second factor was that the reduced numbers in the standard formations of the Atlantic and Mediterranean fleets in comparison with the Grand Fleet allowed much more flexible tactics. A four- or five-ship battle squadron was a very different proposition to Jellicoe's battle line of twenty-four units at Jutland in 1916, and the maximum number of battle cruisers in service from the early 1920s on would be just three. Destroyer flotillas at nine units in the 1920s were half the size of those of 1914–18, and the ships were much bigger and better manned—notably with additional officers, who could help maintain both navigational accuracy and situational awareness to a degree impossible for their predecessors, who usually had been reduced to "following Father"

PHOTO 3.2. Battleship HMS *Emperor of India* during a night-firing exercise in December 1930. At this point she was wearing the flag of Rear Admiral G. F. Hyde, CVO, CBE, RAN. As commander of the Third Battle Squadron, Hyde remains the only Commonwealth naval officer to have served on exchange in seagoing command at flag rank with the Royal Navy. *Sea Power Centre Australia*

(that is, steering in the same direction as the flotilla commander) after any complex encounter.[49]

The environmental conditions that created the opportunity for increased emphasis on surface fighting at night also held potential for aircraft. The first night deck landing on a British carrier came in 1926, with a Blackburn Dart aboard the *Furious*. By 1928, although the doctrinal wisdom was that "The only practicable operations which can be carried out effectively by aircraft in darkness at the present time is bombing against an enemy fleet at anchor, or against shore objectives," there was a rider attached: "This matter, however, is one which is developing so rapidly that it must be kept under frequent review."[50] It did, and it was.

Another element was the establishment of the Tactical School in 1925. This provided not only a center for tactical development, but a mechanism to improve the skills of senior officers. Participation in a tactical course prior to seagoing command soon became customary, the tactical courses having a strong emphasis on group discussion, accepting the inevitable advantages conferred by seniority, yet with reasonably open and free-flowing exchanges of ideas. Its first deputy director noted in later years that "night action received much thought and trial."[51]

By 1925, CB 973, *The Naval War Manual* could describe the Navy's doctrine for main fleet night fighting as:

Night Action.

206. If contact between capital ships has not been made during the day, or if the day action has been indecisive, the Admiral will decide whether or not to seek a night action between capital ships. In making the decision the main factors to be considered are:

(*a*) Danger of losing touch with the enemy and being unable to bring him to action at daylight.

(*b*) Geographical position of the forces.

(*c*) Visibility and weather conditions.

(*d*) Relative strength of the opposing light forces and their tactical position.

(*e*) Enhanced fighting value of light craft at night as compared to that of capital ships.

(*f*) Enhanced value at night of readiness and superior training and technique.

(*g*) Increased liability of chance to influence results.

(*h*) Possibility of surprise and the state of enemy morale as far as it is known or inferred.

If the light forces are able to close the enemy battle fleet at dusk or locate it after dark, they may be able to attack with great effect and produce opportunities for obtaining decisive results in a subsequent attack by capital ships. Early, well organized night attack by light craft on the enemy battle fleet is therefore of paramount importance.[52]

SEEKING THE OFFENSIVE

All these developments provided an opportunity for a step-change in the Royal Navy's attitude to night fighting that itself was part of a deliberate attempt to create a more offensive outlook. A key actor was Admiral Ernle Chatfield, who became commander in chief of the Atlantic Fleet in April 1929 and transferred to command the Mediterranean Fleet the following year. Chatfield, who had been Beatty's flag captain for the entire war, had the benefit not only of complete understanding of the Grand Fleet's experiences, but his employment after achieving flag rank kept him in close touch with both technical and operational developments, including extensive experimental night firings when in command of the Third Cruiser Squadron in 1924. Chatfield believed "that night fighting was to be our great opportunity in the next war. We would surprise the enemy by our efficiency."[53]

It is difficult to work out the exact balance of motivations behind Chatfield's initiative. Although the Royal Navy was becoming aware that both the Americans and the Japanese were practicing firing at ranges of 30,000 yards and more, something of which only the battleships *Nelson* and *Rodney* were then capable in the British fleet, the key exercises that confirmed the inferiority of the Royal Navy in such circumstances of daylight action were not conducted until 1933 and 1934, well after the start of Chatfield's night-fighting drive and after he had become First Sea Lord.[54]

Perhaps with hindsight, in 1944 Admiral William James was unequivocal about Chatfield's motivation: "The British Fleet was relatively no longer so

PHOTO 3.3. Battleship HMS *Rodney* in September 1928. Note the searchlight arrangements and concentrated 6-inch secondary armament in the three turrets, as well as the associated directors. *Courtesy of Drachinifel*

strong in comparison to potential enemies as it had been in previous periods of history, and if efficiency at night battle could be brought to such a pitch that the British Fleet could seek night battle rather than avoiding it, its power to carry out its age-old task of controlling the sea lines of communication would be much enhanced."[55] This may have reflected a view already widely held in the Royal Navy. As early as 1922 in an anonymous *Naval Review* article entitled "Battle by Night," Lieutenant Commander Colin Hutchison stressed that "The fleet of the British Empire will in future be weak in numbers, but numbers alone do not make for victory: it is through the moral factors—boldness, quick decisions and surprise—that great victories are won, and it is the moral factors of which more use must be made in the future."[56] Then–Rear Admiral Drax preached the benefits of close action in the First Battle Squadron of the Mediterranean Fleet in 1929, continuing to make the case for close action—and for night action in later years.[57] In 1932 he wrote, "it is fundamentally wrong that officers should be afraid of, or be taught to avoid, night action. Surely they ought to be glad to engage the enemy on any occasion when the conditions are apparently equal, but may be turned decisively to our advantage by superior skill."[58]

The increasingly ambitious night-encounter exercises would bring their own risks, but these were acknowledged and indeed encouraged by senior commanders. Chatfield's lead was taken up with enthusiasm by his second in command, Vice Admiral William Fisher, and continued when Fisher became commander in chief of the Mediterranean Fleet in 1932. Rear Admiral Andrew Cunningham, commanding the destroyer flotillas in the Mediterranean Fleet from 1934 to 1936, during the second half of Fisher's time as commander in chief, sardonically declared that to make omelettes it was necessary to break eggs. Thus, after the destroyers *Echo* and *Encounter* of the Fifth Flotilla had a glancing collision at twenty-eight knots as they withdrew from a night torpedo attack on the First Cruiser Squadron, the captain of the *Encounter* could report of his subsequent interview with the admiral, "I felt, at once, that I was not being 'carpeted' but that I was receiving very genuine sympathy."[59] Cunningham had his own close call at night when his cruiser flagship found itself approaching a destroyer on the opposite course at a combined speed of thirty-eight knots. The two ships missed each other by a few feet, their upper decks submerged by the combined bow waves.[60] He was not alone. The navigator of the Australian cruiser *Australia* in the Mediterranean in 1935 commented, "During the night exercises I had the usual thrills. Twice I had searchlights put full into my eyes just as the *Berwick* ahead made drastic reductions in speed, and twice we shot past her stern, missing her by inches."[61] By this time, the navy's approach was that described by Admiral Studholme Brownrigg: "The watchword of night action has been dubbed: 'ipomo,' that is the 'instantaneous production of maximum output.'"[62]

Enthusiasm for night fighting did not extend to every senior admiral, however. Then–Vice Admiral Howard Kelly disagreed profoundly with Drax's views in the Mediterranean Fleet in 1929. Kelly came out the worse from exercises to examine Drax's ideas, but this did not stop him from expressing himself freely on the subject.[63] Kelly might not have been in a strong position, having confessed on accepting his appointment as the Mediterranean Fleet's second in command that, "I know nothing, absolutely, of Fleet Tactics, or of handling a Fleet."[64]

For his part, Howard's brother, Admiral John Kelly, whose primary intent as the new commander in chief of the Home Fleet (as the Atlantic Fleet had been renamed) after the Invergordon mutiny of 1931 was restoring discipline and morale, "had no use for Tactics." In a 1933 exercise, when his staff officer for operations briefed him on the plans for destroyer night attacks that would be completed by 2200 that evening, Kelly's response was: "They had bloody well better because I am not going to stop out [stay out] all night for them."[65]

Getting night fighting right required more than just situational awareness and high-risk ship-handling. Admiral Michael Hodges, Chatfield's successor as commander in chief of the Atlantic Fleet, pointed out to the Admiralty in 1930 that, "it cannot be said that our present standard of efficiency is such to warrant engaging in a night action with any great confidence. This will be inevitable while it remains the accepted policy to carry out one night firing only per year."[66] Until well into the 1930s the need to save money continued to restrict the opportunities for live-firings. It would take the Abyssinian crisis for the pursestrings to be opened enough for serials such as "eight ships [the First Cruiser Squadron] at once steaming at high speed and firing at a high speed target towed by a destroyer. What with searchlights, star shells, tracer bullets, recognition and fighting lights, it was an unbelievable sight."[67]

Things could also still go badly awry at night. During Exercise D.H., conducted by the Home Fleet in November 1933, as the Commodore (Destroyers) later explained, "After *Cairo* [a light cruiser] had been sunk, Blue destroyers were searching about in the dark for Red battle fleet when [destroyer] *Versatile* gained touch with what she thought was [the] Red battle fleet, and continued to shadow and report their position until the end of the exercise. Actually *Versatile* was shadowing Blue battle cruisers, with the result that not only were all the remaining Blue destroyers making for the reported position, but Blue battle cruisers also manoeuvred to engage the imaginary Red battle fleet."[68]

THE 1934 MANEUVERS

The Combined Fleet maneuvers of March 1934 have gone down in history as the most public demonstration of the Royal Navy's changed approach to night fighting and its success in developing tactics to meet its demands.[69] They were unusual in that they were conducted in the Atlantic rather than the Mediterranean and in the heavy weather that was to be expected in the Bay of Biscay in early spring. The decision to stage the exercise in what were likely to be difficult conditions seems to have related principally to the associated movements of the Home Fleet and the requirement for fuel economy, even if the easing of restrictions on speed was emphasized in the publicity surrounding the maneuvers.[70]

Within a wider strategic scenario that set out the nature of the conflict between Blue and Red, the Blue (Home) fleet had been tasked with covering a convoy carrying a military force that was to occupy a port on the Iberian peninsula. The selected location would be used as a base for naval forces to interrupt shipping traffic between the United Kingdom (Red) and Gibraltar. The Red (Mediterranean) fleet's mission was to intercept and destroy the convoy.

It was apparent that there were few harbors on the Iberian Atlantic coast that would meet Blue's purposes. In practical terms, this boiled down to two: the entrance to the Tagus River and, farther north, Arosa Bay and Vigo Bay. In reality, a base at the Tagus, barely 250 miles from Gibraltar, would be too open to repeated attacks by Red forces. However, success in the exercise depended upon the safe and timely arrival of the convoy, not on the theoretical utility of the selected destination.

Blue's plan relied upon Red being uncertain which target would be chosen. Thus, the Home Fleet's battle cruisers, aircraft carrier, and supporting units would head for the Tagus as a decoy while the battleships remained with the convoy making for Arosa Bay. The problem, however, was that units of the

MAP 3.1. MARCH 1934 EXERCISE

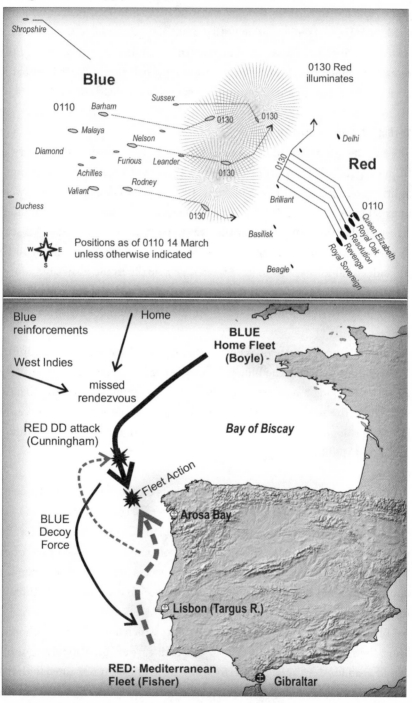

Shropshire

Blue

0110   Barham   Sussex

0130 Red
illuminates

0130   0130

Malaya

Diamond   Nelson

Furious   Leander   Delhi

Achilles   0130   **Red**

Valiant   Rodney

Duchess   0130

Brilliant   0110

N
W   E
S

Positions as of 0110 14 March
unless otherwise indicated

Basilisk

Queen Elizabeth
Royal Oak
Resolution
Revenge
Royal Sovereign

Beagle

Blue
reinforcements   Home   **BLUE
Home Fleet
(Boyle)**

West Indies

missed
rendezvous   *Bay of Biscay*

RED DD attack
(Cunningham)

Fleet Action

**BLUE
Decoy
Force**   ⊕ **Arosa Bay**

⊕ **Lisbon (Targus R.)**

**RED: Mediterranean
Fleet (Fisher)**   ⊕ **Gibraltar**

decoy force were required to achieve a covert rendezvous in the open ocean, since several ships were coming from the West Indies and others from British waters.

Red made dispositions to cover both possible convoy routes, something just practicable if the battle fleet were stripped of its cruisers and destroyers to form a mobile barrier across the lines of approach. Admiral Fisher, the commander in chief of the Mediterranean Fleet, did not plan to depend upon his air arm—a decision that rapidly proved correct. The two fleets encountered an Atlantic depression that manifested itself as a "full gale from the north west." Flying operations were impossible for either side. Then–Vice Admiral William James, commanding the Blue decoy force, later described the waves being "as high as any I had ever seen."[71] The battle cruiser *Renown* had the leading edge of one of her torpedo bulges torn away. The elderly destroyers of the Home Fleet had to run for cover, leaving the big ships denuded of much of their defensive screen, while the decoy force's rendezvous failed. The cruisers coming from the West Indies were nowhere to be found.

Red was not doing much better. The aircraft carrier *Glorious* was badly damaged, and had to write off several aircraft. Various cruisers had to withdraw due to defects, while the destroyers straining to reach their surveillance positions suffered cracking and other structural strains. It was fortunate for Red, therefore, that its submarines had sailed earlier and were in position before the gale hit. It was one of these that sighted and reported Admiral James' decoy force. Apart from assigning a handful of destroyers to shadow it, Fisher did not take the bait and the remainder of Red continued to search for the convoy. This must have been a difficult decision, not helped by the fact that Blue's destroyers had made for the Tagus as their shelter. In the absence of context, their radio traffic was an obvious indication of Blue activity in that location.

A daylight sighting of the Blue battle fleet by the cruiser *Shropshire* gave Fisher the information he needed, particularly since it soon became clear that Blue's battleships did not have any accompanying destroyers. Red's cruisers continued to dog the Blue main body for the remainder of the day. With their speed advantage in the absence of Blue's light forces they were effectively invulnerable.

Any remaining uncertainty on Fisher's part was resolved when the destroyer *Decoy* detected and attacked the sole ship remaining of the small assembly that was intended to represent the invasion convoy. While the Blue commander in chief, Admiral William Boyle, attempted to redispose his remaining units to provide a last-ditch defense, he was soon in trouble.[72] Cunningham's destroyer force attacked just after midnight. What the actual results of its torpedoes would have been is difficult to assess. What was more important for Fisher as he approached with his five battleships was that the "signs of battle," as Cunningham later described them, allowed the commander in chief to adjust his position. Even at this stage, Boyle believed that the Red battle fleet was at least a hundred miles away.

With his battleships deployed on a line-of-bearing to keep their "A" arcs open, Fisher closed on the Blue forces until, at a range of just under seven thousand yards, he ordered simultaneous illumination by star shell and searchlight. The effect was devastating, and there was never any doubt on either side that Fisher would have achieved the complete destruction of the Blue main body with little loss. The defeated commander commented ruefully in his memoirs, "this night exercise showed very clearly the care that he [Fisher] had taken to prepare his fleet for just such an encounter."[73] Fisher's brother, the historian H. A. L. Fisher, was a witness, effusively reporting that, "It is regarded by all the experts as an epoch-making achievement and very daring."[74]

Nevertheless, the extreme weather made for exceptional circumstances in three ways, all of which suggested caution in assessing the results. The first was the absence of many of the destroyers and some of the cruisers. The second was the inability of the aircraft carriers to operate their aircraft (as well as the complete absence of any shore-based aircraft). The third was the effective absence of opportunities for submerged attack. Red's submarines played an important, but limited, reporting role and Blue's submarines had not been able to deploy at all. Thus, although the maneuvers were an "all-weather" event that reminded both the navy and the public that "there are some places other than the English Channel and the North Sea where our Fleets may have to fight," they ended up testing (even if at night) only a single dimension of what had become three-dimensional warfare—on, over, and under the sea.[75]

Admiral Fisher's success in the 1934 maneuvers was not only a morale boost for the Royal Navy. It allowed Ernle Chatfield, First Sea Lord from January 1933, to give his imprimatur to a revision to the Royal Navy's *Battle Instructions* in 1934. The previous edition of 1931 had recognized that night action might be sought between capital ships if the conditions were right.

PHOTO 3.4. The Red Force's HMS *Royal Oak* of the Mediterranean Fleet's First Battle Squadron, taking it green in heavy weather during the 1934 combined fleets exercises. *Sea Power Centre Australia*

The new text emphasized that "night action between heavy ships . . . must be regarded as a definite part of our policy, to be taken advantage of when circumstances require."[76]

INTELLIGENCE ESTIMATES

It remains uncertain just how much other navies knew of the Royal Navy's endeavors. The publicity that the 1934 maneuvers received in the British press must have gotten the attention of the various naval attachés in London, even if did not stir the readers of English newspapers in the countries concerned. The Italian navy seems to have been alerted to British night-fighting training in professional gossip exchanged by the service advisers during the London Naval Conference of 1935. The following year, however, an Italian delegation that visited the new light cruiser *Arethusa* was given the opposite impression, being informed that the Royal Navy was no more keen to fight at night than it had been at Jutland.[77] The Italians, however, were devoting much effort to monitoring British radio traffic and had a much more reliable guide from that. Their 1935 decryptions indicated that the Mediterranean Fleet was conducting intensive maneuvers at night and that, most significantly, these clearly involved its major units.[78]

Despite the Royal Navy's understanding of some aspects of Japanese naval aviation, it was not aware of the extent of the Imperial Japanese Navy's night-fighting development—most notably the 24-inch "Long Lance" torpedo. At a time when the Japanese were seeking the longest possible torpedo ranges, the British Naval Staff expressed concern that the ranges of the 21-inch Mark VII (16,000 yards at 33 knots) and Mark IX (10,500 yards at 35 knots) might be excessive for night fighting, given the evidence in games at the Tactical School that torpedoes were overrunning their intended targets and endangering friendly units.[79] There was cause for this concern, ironically best demonstrated at the Battle of the Sunda Strait in February 1942, when Japanese torpedoes fired at the Australian cruiser *Perth* and the American cruiser USS *Houston* (CA 30) missed the Allied warships and continued into the Japanese force's own anchorage, where they sank or disabled a minesweeper and four transports.[80]

FIGHTING FROM THE AIR AT NIGHT

Nighttime air operations received serious attention as the British carrier force grew. In part this was a direct result of the effort to integrate the Fleet Air Arm into fleet operations in general. The air sorties could wear down the enemy even if the scale of attack possible would not itself achieve decisive results. They could provide long-range reconnaissance and surveillance and spotting information that would enable the big ships to use their guns at their maximum range. Indeed, the idea of naval aviation as a valuable element in daylight battle was quickly expanded into considering it for night action. Many of the skills and communication networks that were required to make things work by day provided the foundation for what was needed at night.

Within a few years of big aircraft carriers converted from battle cruisers entering service, night flying was sufficiently routine that a pilot in the *Glorious* was able to complete more than fifty night deck landings between 1931 and 1933.[81] The carrier's aircraft had accumulated a total of 296 deck landings, compared to 5,659 during the day over much the same period. Some 737 hours had been flown at night and 19,625 by day.[82] By 1939, the Naval Staff could declare, "night operations of all types are perfectly feasible other than that of exceptional darkness or very low visibility. These factors operate primarily against air search and shadowing." Similarly, while in 1931 there were serious problems with long-range over-the-water navigation because of the difficulty of ascertaining wind strength and direction—problems made worse by the fact that employing all-metal aircraft created additional compass errors. Still, these had largely been solved by 1935, and further progress had been made by 1939.[83] Night launching and recovery techniques had been developed with a careful eye on reducing the number of lights that were required to be shown. New procedures to marshal the squadrons in the air and get them on their way without delay had also been introduced.[84]

The Royal Navy never lost its interest in launching nighttime airborne strikes against the enemy in harbor. This formed a part of the contingency planning for war against Italy during the Abyssinian crisis and was exercised against the British fleet in Alexandria. But there was a focus on attacks at sea as well. As Admiral Fisher remarked of his carrier force in February 1936, "The Carriers are ready for war now. War of every kind against ships or shore, day or night."[85]

There was, however, an important constraint, particularly in oceanic conditions. Not only reasonable visibility, but favorable seas were essential for night operations from aircraft carriers. Even in 1939 it was acknowledged that, "With the present deck lighting, however, night deck landing under bad weather conditions, especially when a pilot is tired after flying for some hours, is still too difficult for the average pilot without risk of damage to the aircraft."[86]

By the early 1930s a very high frequency (VHF) homing beacon was under development that would provide aircraft with a bearing to their carrier to use on their return. Its limited range—forty nautical miles—meant there was a low probability of interception by hostile forces, and in any case it would usually require transmission only *after* an airborne strike force had done its work. The Type 72 beacon was tested in the *Courageous* in 1935 and extended to the other aircraft carriers over the next few years.[87] The obvious advantage of the beacon was that the greater navigational certainty it provided meant that the effective range of both reconnaissance and strike aircraft was significantly increased.

The techniques of aerial night search developed progressively. By the mid-1930s, it was possible for a formation of aircraft to find a ship at night and keep the target under surveillance and even illuminated by flares for extended periods. Yet, it was resource-intensive. A mission to find the battleship *Malaya* in the Mediterranean involved seven flare-dropping aircraft spread at five-mile intervals, with six others following two and a half miles astern and at a lower height. This was recognized as "extravagant" in its use of aircraft and practical only when there was a relatively small area to be searched, but its utility for locating an enemy fleet that did not have an accompanying aircraft carrier— the very situation that the Italians faced—was clear. Illuminating targets for surface gunnery also held promise, with ranges of up to eight thousand yards considered practicable, although the aircraft concerned had to pass close over the target for the flares to be effective.[88] The Naval Staff hoped for longer ranges—perhaps up to five miles—but it became clear that larger flares than the type already in service would not provide significantly more light.

Flares also proved useful for illuminating torpedo attacks. In 1933, 811 Squadron developed the technique of deploying flares and making the first torpedo drops about thirty seconds after the flares had lit, followed by a second wave thirty seconds later. This was intended to give the target ships minimal time to react, and it proved effective. On the other hand, during Exercise E.F. in 1934 the surface forces responded by using destroyers and cruisers at high speed to generate wakes that had the appearance of those created by capital ships. This diverted the torpedo bombers to the wrong targets since they proved "unable to identify the types of ships until committed to the attack." Although safety considerations prevented the use of searchlights, the surface forces also believed that pointing them at the aircraft would put the pilots off their aim. Above all, what was clear was that the aircrew needed careful training in target identification and night shadowing in order to be successful.[89] The actual drop of running torpedoes at night was rare, but not unknown, and was certainly exercised to the extent that by the middle of the decade there was real confidence in its capability.

Experience suggested, however, that trying to synchronize air and surface night attacks was too difficult, even though the tactic worked in daylight. By the end of the 1930s, surface forces were enjoined to seize their chance after the airborne attack had gone in. The uncertainties of flight operations and limited endurance of the torpedo bombers meant that they certainly could not wait to be queued by a surface force attack. One tactic favored by 1937 was the use of the "short period at dawn when aircraft can attack from the westward in complete immunity from ship [defense]. The target is clearly silhouetted against the Eastern horizon, whereas the aircraft have the cloak of invisibility given by the night, and attacks can be safely pressed home to ranges where a hit should be certain."[90]

Night operations over the sea were not confined to carrier aviation, although development was more advanced in the RAF's Far East Command than its Coastal Command. The need for a shore-based over-the-water night surveillance and strike capability was evident for the defense of Singapore against a Japanese surprise attack that could extend to carrier-launched air strikes on the British-ruled island or even to outright invasion. The initial deployment of a flight of flying boats to the Far East in 1927 and the formation of 205 Squadron was followed in 1930 by more such machines, with the addition of the torpedo bombers of 36 Squadron; 100 Squadron joined them in 1934. By 1935, the flying boats were undertaking night missions against warships transiting from Hong Kong and occasionally from Australia, as well as against friendly merchant ships. There were repeated efforts to achieve long-range detection and tracking, although the success rate varied considerably, largely due to weather conditions.[91]

The torpedo bombers practiced regular night attacks, although not usually with running weapons.[92] The pace of night flying stepped up with the arrival of Air Commodore Arthur Tedder as air officer commanding in 1936. This included regular flare-dropping exercises against surface targets, as well as much more night formation flying, including formation takeoffs. As with the carrier-borne squadrons, the aircraft also employed flares to assist the torpedo bombers in determining their height for the low-level flying needed for torpedo attacks.[93]

Despite their training programs, the night performance of the RAF units in the combined exercises of February 1938 against the aircraft carrier *Eagle* and her air group simulating a Japanese striking force was not encouraging.[94] The flying boats failed to detect *Eagle* before she launched a strike against Singapore in darkness from 135 nautical miles away—a prospect against which naval commanders had warned the RAF as early as 1934, based on intelligence of Japanese naval aviation capabilities, notably their ability to launch mass strikes from long range.[95]

ALTERNATIVE DETECTION METHODS

The Royal Navy's early success with active sonar meant that passive underwater acoustics received less attention than they did in the German navy.[96] The potential for big ships to detect the "hydrophone effect" of torpedoes (noise generated in water by machinery and movement through the water) in time to take evasive action was a key factor behind the installation of Asdic Type 132 in new cruisers after 1938.[97] Timely torpedo detection had obvious benefits for night action, but there were other advantages as well. In service, destroyers were detected at night in the passive mode at five thousand yards, while three-to-four-thousand-yard detections were achieved in the active mode—often before the ships concerned had been sighted. The advantage of the active mode was that it gave an accurate range as well as a bearing (the passive hydrophone effect could provide only a bearing), but it also meant that the transmitting ship could be detected in turn by passive means.[98] Trials with hydrophones fitted in big ships in the mid-1930s gave some encouraging results, with detection at ranges of up to 110 miles.[99] It was clear, however, that a lot more development was needed to make the equipment fully effective.[100] The invention of radar and the natural priority accorded in sonar development to meeting the submarine threat probably contributed to slowing further development of this detection capability.

LIGHT CRAFT

Reconstitution of a motor torpedo boat (MTB) force in the late 1930s brought with it revival of many of the successful techniques of the CMBs. Exercises rapidly made it clear that if MTBs were appropriately stationed and had an idea of the location of the enemy fleet, they had a remarkably high chance of success. Even in Mediterranean conditions, this was assessed as 30 percent against an unscreened force, with a reduced, but still-reasonable chance against a screened main body, particularly if the MTBs could attack in large numbers.[101]

CONCLUSION: TRANSITION TO WAR

That the main function of night fighting would be associated with fleet action was coming into doubt by the mid-1930s. Although the rules of warfare severely limited the ability of naval units to attack merchant shipping without warning, the strategic situation in the Mediterranean meant that a war with Italy inevitably would involve each side attempting to interfere with each other's sea communications. In Italy's case, its vulnerability lay in the requirement to supply its North African colonies, while the British had to look to the resupply of Malta, itself ideally positioned to act as a base for operations against Italy. Rear Admiral Cunningham wrote in early 1936 to Commander Lord Louis

Mountbatten, captain of the destroyer *Wishart*, suggesting that "our present basis, by day and night, is the attack on what doesn't really exist, an enemy battle fleet. I would keep the best of that, but introduce much more on the subject of what, for a better name, I will call guerrilla tactics for destroyers, and much more on the subject of night fighting and recognition."[102]

In a sense, the revival of the MTB force was recognition of what likely lay ahead for the Royal Navy in war, particularly in the narrow seas of the Mediterranean, around Hong Kong, and in the English Channel. There would be many years of what were much more often "guerrilla tactics" against shipping than operations by opposing battle fleets.

There must be some doubt, however, as to the level of night-fighting expertise maintained by the Royal Navy, even the Mediterranean Fleet, in the years immediately before the outbreak of war. The service was strained by the operational demands of contingencies such as the Spanish Civil War at the same time as it had to expand rapidly. Expertise could deteriorate very rapidly, both because of the high level of training required of every single element but also, as Captain John Godfrey of the battle cruiser *Repulse* stressed, unless there was constant practice, the idea of night fighting became "repellent."[103] Furthermore, notwithstanding the Royal Navy's progress in night-fighting tactics and procedures—as would be demonstrated by the success of Cape Matapan in March 1941—it was still true, as Cunningham himself declared, that "In no other circumstances than in a night action at sea does the fog of war so completely descend to blind one to a true realization of what is happening."[104] Significantly, *Exercises and Operations* for both 1937 and 1938 are both much slimmer volumes with less-ambitious serials, particularly in the main fleets, than in earlier years.

It may have been fortunate for the Royal Navy that radar was coming, but the residual skills and the offensive outlook that had been generated during the interwar period would nevertheless serve it well in night actions in the years ahead.

CHAPTER 4

FORCED TO FIGHT
ITALY, 1940–1943

VINCENT P. O'HARA AND ENRICO CERNUSCHI

A convoy of motorized fishing vessels and small steamers straggled south at five knots against headwinds and rising seas. It carried German reinforcements for the paratroopers who had dropped on Crete the day before. Other than a few small auxiliary minesweepers, the sole escort was *Lupo*, a 670-ton torpedo boat, skippered by the thirty-eight-year-old Lieutenant Commander Francesco Mimbelli of the Regia Marina, the Italian Royal Navy. Mimbelli's aerial escort had departed at nightfall, but after a three-day voyage Cape Spada was only five miles south and the young captain might be forgiven if he considered the perilous part of his journey over. Then, at 2233, three hours after sunset, Mimbelli received a report of intruders to the north. Almost immediately, gunfire erupted. *Lupo* steered toward the commotion and targeted a "large enemy unit" from seven hundred meters. But before Mimbelli could launch, a searchlight pinned the torpedo boat, and *Lupo* started taking hits. She returned fire with all weapons, made smoke, and fired a pair of torpedoes at a cruiser that was rushing toward her. A collision seemed unavoidable. "We passed by only a few meters, I say a few meters only, astern," Mimbelli reported. As *Lupo* drew away, her skipper observed a bright flash. The searchlight disappeared and the enemy fire ceased. Mimbelli assumed that one of his torpedoes had struck home.[1]

In reality, three British light cruisers and four destroyers were attacking the little convoy. The light cruiser HMS *Ajax* reported, "[We] suddenly came across a small Italian destroyer in the mist and smoke of the melee. She was no more than a few hundred yards away and *Ajax* blew her to bits with one six-inch salvo."[2] The British Mediterranean Fleet commander, Admiral Andrew Cunningham, affirmed *Lupo*'s destruction and added that "at least a dozen calques and three small steamers were sunk, shattered or blazing. Some four thousand German troops were left to drown."[3] In fact, *Lupo* escaped, but for two and a half hours the British hunted the loaded transports using radar to find targets in the smoke and darkness. They sank eight of the fourteen transports present, and the death toll was 304. The events of a night action are hard to assess—at the time of battle, afterward, and even eighty years later.

During the first half of the twentieth century a series of general, civil, and colonial wars swept the Mediterranean. In 1914 and again in 1940, seven nations maintained battleship or cruiser fleets on the great inner sea. Chokepoints such as the Dardanelles or the Otranto Straits funneled traffic through the Mediterranean and into narrow subsidiary waters like the Black, Aegean, Adriatic, and Red seas. The through route from Suez to Gibraltar, vital to the British empire's traffic, crossed north-south sea lanes between European powers such as France and Italy and their African possessions, or from the Soviet Union and to the markets for its grain and oil. Narrow waters, crossing routes, vital interests—it was a recipe for conflict. Throughout 1940 the struggle to control these sea lanes and to protect friendly traffic occurred almost exclusively by day. But by 1943 many factors had driven much of the action into the dark and made night combat unavoidable.

During this process Great Britain severely tested the Italian navy's night surface combat doctrine and weapons in a series of battles—most notably the March 1941 Battle of Cape Matapan and the four convoy actions of 12 November 1940, 16 April 1941, 9 November 1941, and 2 December 1942. Over the course of the war, both the fleet and Supermarina, Italy's naval high command, acted vigorously to improve offensive and defensive night combat capabilities and tools. Some of these actions were based on incorrect or partial understanding of lessons learned or were offset by the enemy's own progress in the art of night combat. Even when Italian industry began belatedly to produce critical tools such as very-high-frequency (VHF) voice radio and radar, production was inadequate to equip the entire fleet. Moreover, by 1942 Italian warships lacked fuel oil for routine operations, much less night training. There were also political reasons that impeded Italy's struggle to fight effectively at night. For example, the Axis partners did not routinely share lessons and materiel. Nonetheless, over the course of a hard-fought war, the Italian navy did enhance its skills and improve its results. Even if the Regia Marina never was the equal of the Allied navies at night, it closed ground and reached the point where it could give as well as it got.

BACKSTORY

The history of the modern Italian navy begins with the 1860 unification of the peninsula under the king of Piedmont-Sardinia. Of the preunification navies, only that of Venice (which lost its independence in 1797) had considered night fighting as a way to combat its more numerous, but less-skilled Ottoman foe. In 1717, the Venetian fleet introduced the *Libro d'ordini* (Book of Orders), which included a section for *"Ordini e segnali di notte"* (Night Orders and Signals).[4] Galleys and galleons carried large lights on their high sternposts that used tinted glass to send colored signals. These instructions were first used on 12 June 1717 in the Battle of Imbros, when the Ottoman and Venetian battle squadrons encountered each other three hours before sunset. The fighting continued until 0230 on the 13th, when the Ottoman fleet withdrew up the Dardanelles after losing four ships and suffering three thousand casualties, including prisoners.[5]

The lessons, doctrine, and training for general night combat disappeared from the mainstream of Italian naval thought after Venice lost its independence during the Napoleonic Wars. Piedmont dominated the post-unification navy, and its fleet followed the common doctrine that major warships should avoid night combat at all costs. Night fighting was for light and expendable units only.

THE REGIA MARINA AND NEW TECHNOLOGY

In the 1880s the Italian navy confronted a superior French *Marine Nationale* and remained cautious of its erstwhile enemy, the Austro-Hungarian *k. (u.) k.* Kriegsmarine. Considering itself outnumbered and surrounded, the Regia Marina embraced new technologies such as the torpedo, which gave the smallest platform a way to sink the strongest. Italy formed flotillas of small torpedo boats, led by 840-ton torpedo cruisers, for nocturnal coastal defense. These vessels carried light, one-tube mounts that could be quickly aimed by the crew should a sudden launch opportunity arise. The Regia Marina embraced searchlights, and was the first navy to test radio at sea.[6] But Italy lagged in other areas of night operations. Until 1918, for torpedo fire-control there was only the commander's eyes. Italy used inferior 12×30 commercial-type binoculars that offered high magnification, but gathered less light than the German 7×50 sets

PHOTO 4.1. *Siluro di una Torpediniera*. An Italian torpedo boat's single torpedo mount during the Great War. It was considered the main weapon during night fighting. *Enrico Cernuschi collection*

introduced in 1915. Battleship searchlights, effective out to four thousand meters, provided a defense against enemy torpedo boats, but the smaller searchlights of the flotilla craft reached barely two thousand meters.

Fleet tactical rules that the Italian navy published in 1902 specified that "Nocturnal action is undoubtedly difficult, and will generally be avoided by both sides, unless there are very exceptional circumstances."[7] If a nocturnal fleet action should unavoidably occur, the author recommended starting it with a torpedo attack against the enemy's battleline. He endorsed "precise and exhaustive instructions for all the eventualities of the combat" and stated that "torpedoes are the weapon of choice at night."[8] The Regia Marina also was developing the idea of offensive night operations. In 1912 one of Italy's senior admirals, Paolo Thaon di Revel, advocated surprise attacks against ships in harbor by small torpedo boats or special craft as the primary offensive mission. Another was the interception of enemy traffic.[9] This called for flotilla craft to sail along extended lines of bearing (called a "rake" formation) to find the enemy at night. The contacting ship would fire flares and the flotilla units would concentrate in the target's direction. Ships would launch attacks individually, with aimed, close-range torpedo shots. Gunfire was to be avoided. The pre–World War I torpedo boats and destroyers lacked the compact and simple radio sets required to launch coordinated attacks effectively.

THE GREAT WAR

Between 1915 and 1918, Italy accumulated its Great War night-combat experiences in the Adriatic Sea. During those years ten of the twenty-two surface actions fought between Italian (or Italian and allied forces) and Austro-Hungarian vessels of at least 100 tons were nocturnal. Here is how the Italian navy compared to other First World War navies:

The difficulty of finding the enemy at night can be gauged by this statistic: only nine of twenty-eight Austro-Hungarian night forays against the Otranto Barrage and traffic to Albania made between December 1915 and June 1918 resulted in contact, despite regular Entente patrols, dozens of net tenders, and regular traffic between Italy and Albania.[10] Italy's other night actions generally consisted of fleeting northern Adriatic affairs, sparked by unexpected encounters in narrow waters, followed by sharp maneuvers, sudden outbreaks of gunfire, blinding flashes, and then silent darkness once again. Losses were trivial over the course of a three-year-long naval campaign, and the strategic impact was minor.

Knowing well how hard it was to find the enemy in the dark, Admiral di Revel focused Italy's night offensive activity on attacks against enemy bases by

TABLE 4.1. NIGHT ACTIONS IN WORLD WAR I

NATION	ACTIONS	NIGHT ACTIONS	PERCENT
Italy	22	10	45%
Great Britain	69	19	28%
Austria-Hungary	26	13	50%
France	10	8	80%
Germany	91	27	30%
Russia	50	11	22%
Ottoman Empire	27	5	19%

Note that while the total actions fought is 143, and the total night actions 45, ten actions are counted more than once because there is more than one participant on one side. The detail that supports this table may be found in Vincent P. O'Hara and Leonard Heinz, *Clash of Fleets: Naval Battles of the Great War, 1914–18* (Annapolis, MD: Naval Institute Press, 2017).

special craft or torpedo boats.[11] The Regia Marina conducted nineteen such attacks between 24 May 1915 and 1 November 1918. These sank the modern Austro-Hungarian battleship *Viribus Unitis*, the coastal defense ship *Wien*, and two steamers—results that were good enough to encourage further development of the *insidioso*, or special-attack-warfare concept.[12]

NAVAL POLICY AND DOCTRINE AFTER WORLD WAR I

The postwar Treaty of Versailles left Italy disappointed, and a sense of grievance infected domestic politics. One consequence was the rise of Benito Mussolini's fascist government in 1922. After consolidating power, Mussolini's regime embarked on an expansionist foreign policy that sought to revise the international status quo. In naval terms, this required a fleet capable of dominating the Adriatic and central Mediterranean, of deflecting French threats, and of raiding distant waters. In short, Italy needed a battle fleet, and within the rough confines of the various naval agreements of the 1920s and 1930s, all of which Rome willingly signed, a battle fleet was what Italy built. It reconstructed four dreadnoughts and laid down four new "35,000-ton" treaty battleships, of which three entered service 10 percent to 15 percent overweight, but of high quality and fighting power.

Italian doctrine envisioned various gradations of sea power (*Potere Marittimo*). There was absolute power, in the face of which the enemy would be able to practice sea-denial only; local superiority within a narrower, decisive area, such as Italy practiced during World War II in the Sicilian Straits and Ionian Sea; and *guerriglia navale*, sea-denial used as a potential threat, maintained by expendable forces whose first task was to last as long as possible while forcing the adversary to waste tonnage capacity by resorting to convoys. The Red Sea campaign is an excellent example of *guerriglia navale*.

During the 1930s the Regia Marina's doctrine asserted that sea power could be achieved only by day. To this end, the navy considered daytime superiority delivered by long-range gunfire from battleships and cruisers to be the key ingredient. With respect to combat, Italian doctrine ignored fleet night tactics. For example, in 1940 the fleet had no provisions for using cruiser and battleship main batteries at night. Instead, Regia Marina doctrine stressed the same offensive missions as the navy had assigned to the fleet during World War I —attacks delivered by light craft and special forces against enemy traffic and bases. To support such actions, the navy introduced flashless powder in the

1920s and sought to improve its aiming devices, night optics, searchlights, and illuminating rounds. In 1939 it started the development of secure voice radio. Such policies accorded with the combat experiences of Italy's naval leaders in the 1930s. Admiral Inigo Campioni, fleet commander in 1940, had commanded the destroyer *Ardito* in night combat on 29–30 Sep 1917, and Admiral Domenico Cavagnari, the chief of staff, had commanded the torpedo boat *9 PN* in a nocturnal harbor attack mission and a destroyer flotilla in a 1 July 1918 night battle. They knew such actions were confused, uncertain, and, most of all, inconclusive. Their experience taught that the threat to commercial traffic from out of the night was negligible and they saw no need for specific doctrine to protect shipping against night surface attack.

In these respects, Italian night-fighting doctrine resembled that of Italy's most likely enemy, France, whose navy had a similar technological level. France's Marine Nationale likewise decreed that battleships and cruisers were to avoid the risks of a nocturnal melee and use only their secondary batteries against enemy destroyers with the support of star shells. Destroyers and torpedo boats would undertake offensive patrols in rake formation; torpedo attacks were to be made as opportunity allowed, aiming tubes like a hunter sighting and shooting; and the fleet would sail at night, with cruisers in advance of destroyers.[13]

One problem with Fascist Italy's ambitious naval policy was that the country was the least of the great powers. For twenty-one years, Mussolini's government skillfully acted the part of a world power despite the nation's second-tier economy and limited industrial base. It had fought colonial wars in Libya between 1922 and 1932, Somaliland from 1925 to 1927, and Ethiopia in 1935–36; waged an expeditionary war in Spain in 1936–39; and occupied Albania in 1939.[14] In 1940 the government deposited the military into the middle of a major war with an army that was little better-equipped than in 1915 and still lacked many of the most recent technologies needed, from electronics to aircraft engines. Although war found the navy the best-prepared of Italy's three services, the Regia Marina still had to practice parsimony in its spending and development programs.

Thus, Italian command experience, a big fleet policy, and financial constraints all contributed to Italy's relative disinterest in night combat. The navy felt that night combat requirements could be economically handled by MTBs and torpedo boats, as it had been during World War I. The high command did act to enhance the navy's capabilities, however, by developing, from 1935, a special penetration-and-sneak-attack unit—what today would be considered naval commandos—ultimately called Decima Flottiglia MAS. This required the development of specialized technologies, such as modern scuba gear and underwater, two-man attack craft. The admirals did not ignore other technologies, such as radar, but regarding it as merely one hungry competitor at the limited buffet of Italy's budget, they focused on present needs.[15]

PEOPLE, PLATFORMS, AND TOOLS

Some of Italy's difficulties in night fighting during World War II clearly were rooted in the platforms used and in the weapons and tools that those platforms carried, but personnel issues also played a part. Because the Regia Marina was a big-gun navy, its brightest stars gravitated toward gunnery, not torpedoes. Shortly after World War I one expert commented: ". . . up to now the torpedo has received from us neither the sympathy nor the trust it deserves due to the device's great fragility, and this despite the tremendous services it was called upon to give us."[16]

In 1914 the most modern Italian torpedo boats of the "PN" class (120 tons, 27 knots, 1×57-mm, 3×450-mm TT) had a single central tube and two single side mountings. *Indomito*-class destroyers (1913, 700 tons, 33 knots, 1×120-mm, 3×76-mm, 2×450-mm TT) were originally designed with two central mounts. Contemporary German torpedo boats were carrying four 500-mm tubes and five torpedoes. In 1914 the Regia Marina added a single tube on each side. After 1919 destroyers were uniformly armed with two central triple mounts, although some officers criticized this innovation because with the increased beam of the newer ships it reduced the field of fire. The solution to this architecture problem was torpedo gyro-steering gear, but the diehards remained unconvinced.[17]

Post–World War I improvements included the introduction of Austrian 533-mm torpedoes and German star shells in 1920 and, a few years later, flashless powder for torpedo tubes and guns up to 120-mm. All the new destroyers of the *Freccia* class (1929–33) were equipped with 120-mm/15-cal star shell howitzers. The howitzer's practical range, however, was just four thousand meters. Given that the star shell needed to ignite behind the target to illuminate, this distance was soon considered "*insufficiente*."[18] The navy also ordered a new generation of star shells for 120- and 100-mm guns, while the modernized *Cavour*-class battleships and the heavy and modern light cruisers received an

APG *notturna* (*Apparecchio di Punteria Generale* or general aiming device) on each wing of the bridge. The APG was an electro-mechanical apparatus with a 1.5-meter stereoscopic rangefinder (whose optics lacked an antireflective treatment such as that adopted by the Germans after 1935). This fed range and bearing data to the ship's secondary director, which was electrically linked to the follow-the-pointer gun turrets and mounts. Modern destroyers from the *Sella* class on, along with the modern torpedo boats, all carried this system, which the navy considered adequate for the job.[19]

The story of the thirty-two torpedo boats of the *Spica* class is especially relevant. The 1930 Treaty of London did not limit torpedo boats that displaced less than 600 tons. Consequently, Italy built this type in large numbers, not only because it could, but because the navy and the IGM (*Istituto di Guerra Marittima*, or Naval War Institute) had identified a need "for small fast torpedo-armed units with the idea of using them for night attacks on enemy formations during torpedo melees, which the IGM saw as possible and even desirable."[20] The prototypes of that class, the torpedo boats *Spica* and *Astore*, ordered in 1931, had a 450-mm single mounting on each side and a central twin mounting (three-tube salvo); units built the next year had four single mounts (two-tube salvos).

PHOTO 4.2. An Italian Astramar binocular 12–20–40×80 fabricated by San Giorgio of Genoa on the cruiser *Abruzzi* in July 1940. *Enrico Cernuschi collection*

Admiral Wladimiro Pini, deputy chief of staff from 1932 to 1938, championed the two-tube broadside. He argued that better fire-control computers and improved torpedoes had eliminated the need for large, wasteful salvos, and that single mounts, being more dispersed, were less subject to battle damage. Exercises during the 1930s seemed to affirm Admiral Pini's views and taught that at night efficient communications with the bridge were key. There, a torpedo transmitting station on each bridge wing supplied the rate of change, while the officer responsible for pressing the firing circuit visually estimated enemy inclination and speed.[21]

In mid-1938 the Regia Marina increased the intensity and realism of its night-combat training. The 1935 Ethiopian crisis had established the British Royal Navy as a potential future foe, and intelligence that the British were conducting night-combat exercises in the Mediterranean provided motivation. The new exercises led Admiral Cavagnari and the deputy chief of staff, Admiral Campioni, to conclude that torpedo boats needed two central twin mounts. The improvement was introduced on the last four ships of the class—*Alcione, Aretusa, Ariel*, and *Airone*—and retrofitted (from early 1942) on surviving vessels. In many of the actions fought in 1940–1941, however, the Italian torpedo boats could only launch a two-torpedo salvo. Cavagnari considered the *Spica*s too large in any case and believed that the *mototorpediniera*—a 62-ton motor-torpedo boat that was more seaworthy than the 24-ton MAS (*Motoscafo Armato Silurante* or torpedo-armed motorboat) boats of the "500" class—to be a better option for night combat. However, *Stefano Turr*, a 1936 experimental 52-ton boat that carried the same torpedo armament of a *Spica*-class torpedo boat, was a failure because her diesel engines never functioned satisfactorily despite almost three years of testing and tinkering. The large MTB was an example of an improved weapon platform that was not available when the war started, once again due to an absence of priority and the availability of an acceptable alternative. The Regia Marina's MAS boats, while clearly limited, at least had a record of success in World War I and had been improved since then. Once war started, however, the small MAS boats proved unsuitable for most operations in the open Mediterranean. Over the entire war they achieved just three successful attacks out of ten launched at night, and in most of their forays they were not able to find the enemy at all.

LAST-MINUTE ACTIONS AND POLICY

In conjunction with better training, the Italian navy improved its night combat tools. Its actions included

- ▸ Developing a new longer-range 120-mm/17 star shell howitzer.
- ▸ Equipping the *Zara*-class heavy cruisers with two 120-mm/15 howitzers and the modern battleships with four old 120-mm /40 guns for star shells, despite their slow rate of fire and inadequate range.
- ▸ Beginning development of a 152-mm star shell with a range of 14,000 meters.[22]
- ▸ Distributing Italian-made 7×50 binoculars, copied from German models.
- ▸ Introducing a mounted 12–20–40×80 trifocal binocular.[23]

The Italian naval staff also prioritized communication technology. From early 1940 the navy electronic laboratory of Livorno (Marinelettro) focused on perfecting the TPA (Telefono per Ammiragli), a 50–80-cm directional, one-channel VHF radio system, similar to the U.S. Navy TBS, whose range was thought to be line-of-sight and thus interception-proof. This would allow secure and rapid tactical communications, one of the major keys to successful combat. However, given Marinelettro's small staff (fewer than thirty) and limited budget, the development of this radio delayed work on shipborne radar.[24]

To summarize: in the late 1930s Italy's naval policymakers believed that a careful application of Potere Marittimo, founded on control of the central Mediterranean, and bluff could continue to deliver positive results in political confrontations with Britain such as occurred over Ethiopia in October 1935 or with Spain in January 1937. Control of the central Mediterranean in this scenario envisioned cutting British traffic through the Sicilian Narrows and mounting a strong defense of the long Italian coasts. By late 1942, after all eight Italian battleships were in commission, victory in a shooting war also would be possible. Mussolini's regime envisioned that it would fight such a war as Germany's partner, and that the war would be brief, resulting in a favorable compromise peace.

The two general scenarios envisioned for night-combat included chance encounters by the fleet and offensive missions by light units. In short, Italy planned to fight night actions when it wanted and to avoid them otherwise. Acting offensively, and also in accordance with the concept of Guerriglia

PHOTO 4.3. Crown Prince Umberto on the starboard flying bridge of the battleship *Vittorio Veneto*, Naples, 4 December 1940. The mechanical calculator is a *colonnina*, that is, the APG (*Apparecchio di Punteria Generale*) directing, in accordance with the gunnery officer's visual range estimation, 152-mm and 90-mm gunfire at night. Visible behind the sailors are the gyro-stabilized basements of three of the five lookout optical stations and the fire-control station of the four axial-stabilized director system for the 152-mm starboard battery. *Enrico Cernuschi collection*

navale, all MAS, torpedo boats, and destroyers were to use stealth and the torpedo as their main weapon. Ships were to resort to guns and searchlights only after they themselves had come under gunfire. The night cruising rules for the fleet specified that destroyers were to follow the cruisers, giving the larger ships the first sighting opportunity. If an enemy were sighted, the cruisers would immediately turn in the opposite direction to withdraw from the zone of action as the destroyers dashed forward to attack with torpedoes. The navy's combat rules had no provisions for defending convoys against enemy night attacks.

WORLD WAR II

From June 1940 through the armistice of 8 September 1943, the Italians fought twenty-three night actions with major warships (torpedo boats and larger) against major enemy warships. Italy instigated nine of these, the British started twelve, and two were accidental encounters. Up through April 1941 there were thirteen actions, with nine instigated by Italy, all interceptions of

Allied traffic by torpedo boats. From May 1941 through June 1943, there were ten night actions, one accidental and nine instigated by the British against Italian traffic. Thus, during the war, two-thirds of Italy's surface actions were fought at night (not including MTB actions, which were nearly all nocturnal). This did not match Italian expectations that the navy would fight offensive night battles against enemy traffic and avoid the rest. In fact, Italy fought nine actions against enemy traffic and ten to protect friendly traffic—a category of action that had been considered unlikely before the war and thus was not adequately addressed in doctrine. Even more surprising was the only "fleet" action, the Battle of Cape Matapan of March 1941. It was the most famous night battle Italy fought, and the most significant in terms of its impact on Italian doctrine, but in fact it was an anomaly. Fleet squadrons never again met at night, and on several occasions when a day action could have continued into the night—in both the First and Second Battles of Sirte, in December 1941 and March 1942—the Italians broke contact after dusk, being aware of the enemy's radar advantage.

Italy instigated five night actions during the Parallel War period of June 1940 through January 1941, when it fought without German assistance. Three of these were in the Sicilian Channel and the others were in the Red Sea and Aegean. In these, Italy lost four flotilla craft and saw another three significantly damaged; in return, Italian vessels damaged an enemy light cruiser and a tanker. In part this unfavorable ratio arose because the torpedo boats delivered small salvos of small torpedoes, while the MAS boats, which on paper carried a torpedo boat's punch in a penny package, proved too susceptible to sea and weather to operate efficiently in anything other than perfect conditions. Italian doctrine, based on the "shoot-and-scoot" school of torpedo warfare, called for aimed shots from close distances, followed by disengagement. Ships acted independently. In three cases the escort's strength discouraged the attacker, and in one instance four torpedo boats made contact, one after the other, but only one actually managed to get two torpedoes into the water. Night gunnery was discouraged.

FIRST ACTION, 12 OCTOBER 1940
Through the summer and into the fall of 1940 Italy's offensive night-combat doctrine was untested. At the end of August, British forces transited the Sicilian Straits for the first time, passing undetected, assisted by bad weather that

kept the MAS boats in port while a pair of torpedo boats raked the wrong strip of ocean. In the next operation a convoy from Alexandria, escorted by the Mediterranean Fleet, including four battleships and two aircraft carriers, arrived in Malta on 11 October. At 1100 on the 11th an Italian airliner spotted British warships a hundred miles southeast of the island. Additional air searches found nothing, so the navy made generic dispositions based on the limited information that it possessed. The 1st Torpedo Boat Squadron (*Airone, Alcione, Ariel*) and the 11th Destroyer Squadron (*Artigliere, Geniere, Camicia Nera, Aviere*) departed Augusta starting at 2000 on the 11th. They proceeded 120 miles southeast, formed a long line-of-bearing and began to sweep west toward Malta. From north to south the squadrons could cover a thirty-two-mile-wide ribbon of ocean. Meanwhile the 7th Destroyer Squadron and a pair of MAS boats from Pantelleria patrolled between Marettimo Island and Cape Bon in case the British were headed for Gibraltar. Two MAS boats took an ambush position off Malta while a pair of light cruisers raised steam at Palermo. Supermarina also suspended traffic to Africa.

Three merchant ships escorted by a light cruiser and two destroyers departed Malta for Alexandria at 2230 on the 11th while the fleet loitered fifty miles to the south, and the light cruisers *Orion* and *Ajax* scouted seventy miles northeast by east of the fleet. Other units were positioned as shown in Map 4.1. The cruisers had turned northeast at 2200 on the 11th, *Orion* on a course of 030 degrees and *Ajax* of 060 degrees. The Italian rake approached on a crossing course and at 0137 an Italian torpedo boat, *Alcione*, spotted *Ajax* an estimated 18,000 meters off her port bow—an extremely long-range sighting made possible because the moon, three days short of full, was sinking behind the enemy ship. Up to this point it was a textbook case of applied doctrine, and the Italians held every apparent advantage.[25]

Alcione radioed her sighting and turned southwest toward *Ajax*. At 0142 *Airone* and *Ariel* spotted *Ajax*. At 0148 the squadron commander ordered the converging torpedo boats to attack. *Ajax*'s commander, Captain E. D. B. McCarthy wrote that at 0155 his ship had sighted a destroyer off her starboard bow and he quickly determined that she was an enemy and passed the order to open fire. Each side claimed the other side flashed recognition lights.

Meanwhile, *Alcione* approached undetected broad on *Ajax*'s port beam and at 0157 launched two torpedoes from eighteen hundred meters. Since *Airone* had briefly fouled her range, she continued and uncorked her other

MAP 4.1. 11–12 OCTOBER 1940 ACTION

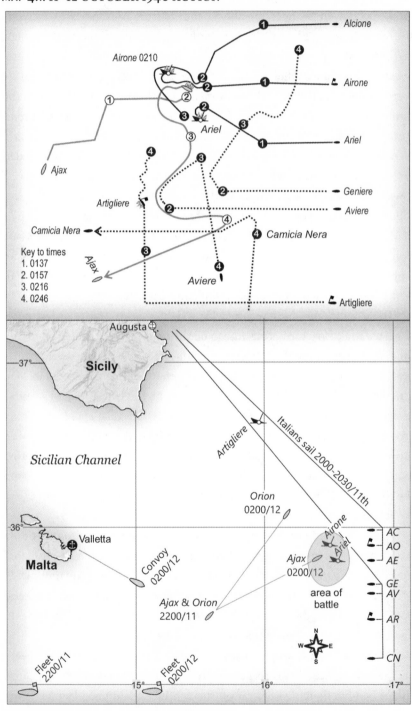

Alcione

Airone 0210

Airone

Ariel

Ariel

Ajax

Geniere

Artigliere

Aviere

Camicia Nera

Camicia Nera

Key to times
1. 0137
2. 0157
3. 0216
4. 0246

Ajax

Ajax

Aviere

Artigliere

Augusta

37°

Sicily

Sicilian Channel

Artigliere

Italians sail 2000-2030/11th

Orion
0200/12

36°

Valletta

Malta

Convoy
0200/12

Ajax
0200/12

Airone

Ariel

Ajax & Orion
2200/11

area of
battle

AC

AO

AE

GE

AV

AR

CN

N
W E
S

Fleet
2200/11

Fleet
0200/12

15°

16°

17°

two at 1058. Within seconds, *Airone* launched two torpedoes off the *Ajax*'s port bow, estimated range eighteen hundred meters, and *Ariel* another off the starboard bow from a thousand meters. *Ariel* got only one weapon in the water before a flurry of shells punched into her bridge "and interrupted transmission of the commander's order."[26] This would suggest that *Ariel* did not begin launching until at least two minutes after *Airone*. In any case, seven torpedoes sped toward *Ajax* from three different directions.

"Either just before or immediately after fire was opened two hits were received," McCarthy wrote. *Airone*, meanwhile, reported that having "completed the launch of all torpedoes, opened fire on the enemy one minute before [the enemy] began shooting." If *Airone*'s chronology is accepted, she opened fire at 0158–0159 and *Ajax* at 0159–0200.[27]

As *Airone*'s second pair of torpedoes hit the water, her 100-mm guns snapped off four quick broadsides and landed two shells on *Ajax*'s bridge, one in the radar room, disabling the Type 279 receiver. McCarthy's report does not mention any maneuvers until "several minutes after fire had been opened a second destroyer was sighted on the port bow . . . immediately . . . speed was increased to full and the wheel put hard to starboard, with the double object of avoiding torpedo fire" and opening firing arcs for the secondary battery.[28] The Italian accounts are unanimous that their torpedo attack was frustrated "by the counter maneuver of the enemy cruiser [*Ajax*], which immediately reversed course."[29] The Italian ships all estimated the cruiser's speed and heading differently, which is cause to question the accuracy of their aim.

In response to the sudden attack, *Ajax*'s forward guns targeted *Ariel* to starboard, estimated range four thousand yards, interrupting her torpedo attack as related, and devastating her engine spaces. This was good shooting. Apparently *Ajax*'s estimate of range was better than *Ariel*'s. Subjected to a sustained and accurate bombardment for several minutes, *Ariel* sank rapidly, the captain and senior officers going down with the ship.

After what McCarthy reported as "a few minutes" *Ajax* spotted *Airone* to port and made a sharp turn to starboard. At the same time another 100-mm round, again from *Airone*, penetrated the cruiser's hull six feet above the waterline igniting a blaze that burned for three and a half hours. *Ajax*'s two twin 4-inch portside mounts answered. *Alcione* also engaged within seconds of *Airone* and fired fifteen salvos, all missing although she claimed several hits. At the same time, *Alcione* could not obtain a satisfactory firing solution for her last two torpedoes, so she drew away intending to line up for another attempt. *Airone* was not so fortunate. Her captain, also thinking to disengage, ordered smoke, but *Ajax* swung to port astern of *Airone* and came up on a parallel course. A broadside of 112-pound shells pummeled the torpedo boat's two aft 100-mm mounts and sparked a large fire amidships. Over the next few minutes, other shells smashed *Airone*'s rudder and engine room as the cruiser made several runs up to and past the drifting torpedo boat, once closing to point-blank range and hosing the smaller vessel with machine-gun fire.[30]

The pyrotechnics drew the 11th Destroyer Squadron to the scene. *Aviere* sighted *Ajax* first at 0210. At this point, the British cruiser had gained the advantage of light, which only enhanced her already effective marksmanship. As *Aviere* closed and maneuvered to launch torpedoes from a suitable angle, a 6-inch shell smashed her aft 120-mm mount and another penetrated the hull, causing major flooding forward. *Aviere* answered at 0215, turned east, and broke contact by 0218. She hit the cruiser twice on the port side. The first round disabled a 4-inch mount and the other struck *Ajax*'s side armor and sent splinters flying. The metal fragments "tore the [torpedo] warhead . . . practically in half, scattering T.N.T. on the deck." *Geniere*, the northernmost Italian destroyer, glimpsed *Ajax* at 0218. "The sighting took place in conditions that, due to distance and angle, did not allow launch. As soon as the attack maneuver began, *Geniere* lost sight . . . and therefore was unable to engage."[31]

A second destroyer, *Artigliere*, turned north at 0200. After a half-hour, she was surprised by flashes, and then geysers erupted around her as *Ajax*, approaching from the opposite direction, opened fire from just a few thousand yards. *Artigliere* released one torpedo before a 6-inch round barreled into her bridge and killed the squadron commander. Based on the destroyer's report of an explosion at the expected time, Maristat (the Italian naval staff) credited *Artigliere* with a torpedo hit against the enemy, but in fact she missed.[32] Not so British gunfire. A 6-inch shell detonated *Artigliere*'s forward 120-mm ready ammunition, igniting a large blaze. As *Ajax* steamed past, the destroyer scored with two 52-pound shells that hulled the cruiser on the starboard side and resulted in some easily isolated flooding. Then three rounds rattled into the destroyer's forward engine room, and a fourth demolished her central boiler. At 0245 as *Artigliere* lost way, *Ajax* fired two torpedoes a thousand yards back at the burning ship and then turned east. Despite the close range and drifting target, both torpedoes missed.

Camicia Nera, the southernmost destroyer, sighted *Ajax* at 0246, five thousand yards to the northwest. She and the British cruiser swapped a few ineffective salvos. But *Ajax* lacked flashless gunpowder, and repeated broadsides had ruined her crew's night vision. The British cruiser misidentified her enemy as a cruiser, and when this adversary disappeared in what seemed to be a smoke screen, *Ajax* turned southwest to break away; the fleet commander, Admiral Cunningham, had ordered her to retire toward the battleships "in the belief that the enemy's whole fleet might be in the vicinity."[33] *Camicia Nera* crossed *Ajax*'s wake on a diverging course and lost contact.

Ajax suffered thirteen killed and twenty-two wounded in this action. She expended 490 six-inch shells, 75 four-inch shells, and 4 torpedoes. She would return to action after three weeks in dock. *Artigliere*'s crew controlled her fires, and at 0500 *Camicia Nere* took her in tow, but at 0840 other British cruisers caught up and sank the crippled ship. Italian losses were thus two torpedo boats and one destroyer sunk and one destroyer damaged.

Until the shooting started, the action seemed to affirm Italian doctrine. The torpedo boats conducted a textbook approach, but only put seven (of twelve available) torpedoes into the water. In contrast, British doctrine, as expressed by the Mediterranean Fleet's vice admiral, light forces, was that "all available torpedoes should be fired when a suitable target presents itself."[34] The 11th Destroyer Squadron, which possessed greater torpedo capability, never attempted to concentrate; instead, the ships presented themselves individually to *Ajax*'s gunners. On the other hand, Italian flashless propellant worked well. Gunfire was completely invisible to the British, according to their accounts, while McCarthy's report bemoaned the negative effect that gun flashes had on his crew's vision.

What lessons did the Regia Marina gleam from this action? With the facts laid out it seems clear that torpedoes should be used whenever possible and in the greatest quantity, even if conditions were not perfect, rather than being doled out in prescribed measures at exactly the right distance, as if each weapon were precious. However, after examining the reports, Maristat—whose officers had followed a policy of parsimony throughout their career and who now faced the need to fight a war with limited resources, where indeed each torpedo was precious—reached a different conclusion: that its ships had faced and damaged at least two enemy cruisers and had torpedoed at least one. From this, Maristat concluded that its ships had been too aggressive with their gunfire and this explained the severe losses. Supermarina quickly issued directives stating that the flotillas needed to emphasize stealth and rely upon torpedoes. They were to shoot only when (and if) sighted by the enemy. But this was only a reminder, not a change in doctrine based upon a lesson learned. The fate of the 11th Squadron as its units closed individually to attack should have rung alarm bells. Command and control problems were apparent at the time, even if the relatively poor marksmanship was not. What went right from the Italian point of view? These were the best-trained squadrons, and the naval staff considered their efforts to be excellent. The rules for setting up combat seemed to work perfectly. Yes, losses were harsh but at least, so the Italians thought, they had inflicted punishing damage in return. Thus, the Regia Marina continued to conduct offensive night actions using rake formations and opportunistic torpedo attacks—just more cautiously.

On 28 November four *Spica*-class boats intercepted a heavily escorted Malta convoy at night in the Sicilian Narrows. However, they only fired two torpedoes of the sixteen that they carried. Conditions were difficult, but it was more a case of excessive caution. Things went differently on 31 January 1941, when two torpedo boats based at Leros conducted a two-vessel rake off the north coast of Crete and intercepted a tanker and a steamer escorted by the light cruiser *Calcutta* and a sloop that together were a detached portion of a larger British convoy. Once again, everything worked by the book. The Italians approached undetected. At 0240 the Italian torpedo boat *Lince* launched two torpedoes from about a thousand meters and *Lupo* two more forty minutes later at eight hundred to nine hundred meters. Both claimed hits, and in fact one of *Lupo*'s weapons seriously damaged the tanker *Desmoulea* (8,120 GRT). *Lupo*'s captain, Lieutenant Commander Mimbelli, noted: "During the action I did not consider it appropriate to use my guns because I did not want the escort to ascertain my position with greater accuracy and engage me with their bigger guns."[35]

Maristat regarded this action as proof that the night offensive doctrine was effective if properly applied. *Lupo*'s report identified the need for a greater torpedo broadside and indeed, the boats were being refitted, but the platform could not be improved very much. The true revolution in Italian naval night-fighting policy only occurred in the spring of 1941 after the Battle of Matapan.

MATAPAN

In the night action at Matapan, the Mediterranean Fleet's three battleships surprised a pair of Italian heavy cruisers. The British sighted the enemy first from four thousand yards, and once that happened no amount of training, improved flares, specialized ammunition, better optics, night fire-control, or gunnery capabilities would have altered the outcome. The only things that would have saved the Italian squadron would have been a first sighting, followed by immediate flight or an effective surprise torpedo strike.

Based on reports from an earlier air attack, the British were expecting to find a stopped battleship. In fact, their prey was a heavy cruiser and they were steering toward it on the basis of a contact on HMS *Valiant's* Type 279 radar (the other two battleships still did not carry radar). The Italians were approaching to rescue their division mate and neither squadron expected the other's presence. Once they made contact Cunningham maneuvered, using short-range radio, and the battleships opened fire from 2,900 yards, surprising the Italians and scoring first salvo hits.

Matapan was significant because it forced the Italian navy to accept that a nocturnal capital ship action instigated by a superior fleet was a real danger. Once the 1st Cruiser Division was caught by surprise, its slaughter was a foregone conclusion. Destroyers, however, proved even less effective than in the 12 October action, inflicting zero damage at the cost of two ships sunk and one damaged. The lesson that the Italian high command learned from this disaster was that a fleet commander could exercise command and control at night, so it must be a matter of the proper tools and tactics. The lesson that continued to elude was that the navy's torpedo doctrine was ineffective. The battle alerted the Italians to the existence of operational shipborne radar, thanks first to intercepted British signals and then from the Germans, who also shared that they themselves had possessed shipborne radar for nearly four years.

Italian fleet operations were governed by instructions issued by the fleet commander, in this case Vice Admiral Angelo Iachino, titled: *Norme di massima per l'impiego in Guerra* (or *General Rules for Conduct in War*). The rules covered formations, navigation, maneuvers, use of weapons, port security, and use of aircraft. They were comprehensive—perhaps excessively so. For example, Chapter 6, Article 3 (of 17) Subsection A, *"Premessa dul servizio sicurezza,"* or "Outline of Security Duties," instructed that, "if in port and the air alarm sounds, hatches should be closed, or if it is night, they should be verified as

closed." Such rules gave solid guidance in situations anticipated during peacetime or experienced in war up to date, but they did not promote initiative.[36]

In April 1941, jolted by the shock of Matapan, Iachino began drastically
revising the general rules to provide for fleet combat at night. He tested and
refined these revisions during exercises conducted in May and June 1941. In
July 1941 he published a revised set of general rules. These held that the main
task of the escort in a night battle was to prevent surprise, rather than to stay
back and act as a counterattack force during the withdrawal of the main force.
The new rules expanded the number of formations that were available for nocturnal navigation and refined the conditions for use of weapons from 381-mm
("All guns loaded and elevated to maximum night visibility range") down to
20-mm ("ready, as during day").[37] Other important improvements made by 22
July included

> ▸ More night combat training. This program lasted until September 1941,
> when the growing fuel crisis forced cutbacks.
> ▸ Putting destroyers four thousand meters ahead of the fleet in the
> direction of a probable enemy contact.
> ▸ Initiating combat would be accepted at night if the enemy were of
> similar strength and would be engaged with all weapons, torpedoes
> being secondary to gunnery.

Better tools took more time. These included

> ▸ Gyrostabilized searchlights.
> ▸ Improvements in the APG and its use to direct main batteries as well as
> secondary ones.
> ▸ Improvements to night optics, including blinders and antiglare and
> antireflective treatments.
> ▸ Development of a *grande gittata G.G.* (long-range) star shell for the new
> generation of 135-mm, 120-mm, 100-mm, and 90-mm guns. However,
> only the two smaller shells reached the fleet between 1941 and 1942.

The shift toward accepting night combat if the odds were nearly equal
had other drivers beside Matapan. Effective dive- and torpedo-bombing had
persuaded naval staff officers that ships were safer from the air threat while
cruising at night and that night might become the preferred operating environment—at least when sailing beyond the range of friendly fighters.

Given the navy's focus on gunnery, it is unremarkable that Iachino devoted attention and care to night-firing arrangements. Based on a study of combat experience and findings from exercises, fleet headquarters generated a flood of memoranda and directives, covering subjects such as elevation and training of turrets (3 June 1941); effective night ranges (30 June); star shell ranges (5 July); glare protection (12 July); searchlight effective ranges (5 July); APG shortcomings (4 October); and the heights of water columns for spotting (25 December). These were all important matters, but it is difficult to disagree with the Maristat officer who wrote on 19 May 1941 that "The problem (of night combat) is grave. We do not hide it. . . . We should have done these things in peacetime, but it is useless to dwell on the past."[38]

A basic assumption behind the promulgation of detailed instructions is that they can be followed. Iachino's rules were full of admonitions about the importance of "clarity" and the need to quickly determine the enemy's general strength, location, and direction of advance. They were silent about what to do in situations of complete confusion.

PHOTO 4.4. The German DeTe radar on the Italian destroyer *Legionario* during summer 1942. The Germans sold Italy a few naval radar sets in April 1941, but the first wasn't delivered until May 1942, and it was fitted only on board *Legionario*. In June 1941 Italian dictator Benito Mussolini purportedly said that after the war he would build some large cemeteries and the largest would be reserved for German promises. *Rivista Marittima*

In addition to revising doctrine, the Italians progressed in technical matters. The navy gave new priority to the development of radar, and in May 1941 the Regia Marina conducted its first test of radar afloat with the 70-cm EC 3bis set on board torpedo boat *Carini. Littorio* received an EC 3bis that August. However, it proved unsatisfactory in the way it displayed the return echoes.[39] The set's emissions were returned aurally, like sonar, and required a skilled and attentive operator. It was useless for gunnery. The first visual displays were little better. Power and reliability problems further delayed the introduction of a usable version, the EC 3ter Gufo (Owl) until September 1942. The Regia Marina could not wait that long, and in May 1942 it installed its first efficient radar—a German FuMO 24/40 G (DeTe)—on destroyer *Legionario*. By September 1943 nineteen Italian warships, from battleships to torpedo boats, carried Italian or German units. Gufo sets were basically handcrafted; production reached four units a month by mid-1943.[40]

By the autumn of 1941 the Italian navy was optimistic that it had overcome its inferiority in fighting, or had at least narrowed most of the gap revealed by Matapan and the earlier actions. However, from that point forward, British warships increasingly confronted Italian traffic at night, presenting the Regia Marina with a new problem that it had not prepared to face.

IN DEFENSE OF TRAFFIC

Allied surface forces first attacked an Italian convoy on the night of 12 November 1940 in the Strait of Otranto. Two light cruisers and a pair of large destroyers, operating in conjunction with the carrier raid on Taranto, encountered a four-ship convoy escorted by an old torpedo boat, *Fabrizi*, and an auxiliary cruiser. The sighting was mutual. *Fabrizi* charged, making smoke, and tried to give the merchantmen a chance to scatter. Her attack failed, however, when she came under effective fire before she could close to prescribed torpedo range. After shouldering the escort aside, the British squadron destroyed the convoy, sending four ships grossing almost 17,000 GRT to the bottom and leaving *Fabrizi* badly damaged.[41]

The Italians regarded this event as an anomaly—and with reason. Despite further attempts by British and Greek destroyers, none of the daily convoys between Italy and Albania was ever again attacked by surface forces. In fact, even on the exposed routes to Africa more than five months elapsed before another Italian convoy suffered a night surface attack. Once again, however, the results were catastrophic. The setback occurred on 16 April 1941 off the Tunisian city of Sfax, when four large British destroyers, equipped with new Type 286 radars, intercepted five merchantmen grossing 14,498 GRT, which were escorted by three Italian destroyers. This time, the British surprised the convoy, opening fire from a few thousand yards astern with the advantage of the light. In the confused action that followed, the Italian escorts dashed in to deliver uncoordinated torpedo attacks. British guns hit repeatedly, and Italian fire missed. The end result was that all the merchantmen and two destroyers were sunk and the survivor was heavily damaged. The British lost the destroyer HMS *Mohawk* to two Italian torpedoes. It is telling, however, that a fatally damaged ship delivered this blow and that the torpedoes were fired on a midshipman's initiative. What was the message that Supermarina gleaned from this event? It was another anomaly. Except this time it was not. It was the result of intelligence.

Intelligence can be a vital aid in night combat. In this case, signals intelligence (SIGINT) enabled the interception by putting the British flotilla in the convoy's general vicinity. Once the British had a source of intelligence that could regularly warn of convoy sailings, the entire dynamic of combat in the Mediterranean changed. Now, using this intelligence, the British could effectively dedicate a cruiser-destroyer strike force—teamed, after September 1941, with radar-equipped aircraft—to prey on Italian traffic.

This new dynamic was forcefully expressed on the night of 9 November 1941, when the strike force fell upon an unusually large convoy escorted by a division of Italian heavy cruisers and two destroyer squadrons. The heavy cruisers were there to counter British cruisers, but they expected to fight during the day due to the abiding belief that the probability of a night interception by surface forces was near zero unless the enemy knew the convoy's exact course and speed. Supermarina believed the most likely threats were night torpedo strikes from aircraft or submarines. The cruiser commander, Rear Admiral Bruno Brivonesi, had been on his job for seven months. He had escorted important convoys and believed that confusion and a loss of command and control were the greatest dangers at night. He believed that the heavy guns of his cruisers were a "pure danger" (to his own units) at night and positioned the six ships of the support group on the convoy's starboard quarter to maintain separation. The close escort of six destroyers formed a protective screen with one fore and aft and two on each side of the merchant vessels.[42]

An aircraft sighted the convoy, and the British task force, Force K, comprising two light cruisers and two destroyers, put to sea. Although Malta had assigned an ASV (Radar, Air to Surface Vessel) aircraft to guide Force K to its target, the plane's radar had malfunctioned, and SIGINT information about the convoy arrived in Malta only after the attack started. Nonetheless, as Force K was approaching its turnback point, it spotted the convoy in bright moonlight. It was another fortuitous interception. The senior officer, Captain William Agnew, had fought in a small action against a German convoy that had featured confusion, collisions, and the convoy's escape.[43] For this assignment he devised a simple but effective doctrine that had two basic rules: Ships would fight in line-ahead formation. That is, they would follow the leader to facilitate command and control. This would also "avoid recognition problems and give freedom of torpedo fire." Ships would engage enemy escorts first, or, if they were already engaged with the convoy, would shift fire to any new escorts that might appear. Ships would try to keep escorts fine on the bow until

MAP 4.2. BETA CONVOY ATTACK, 8–9 NOVEMBER 1941

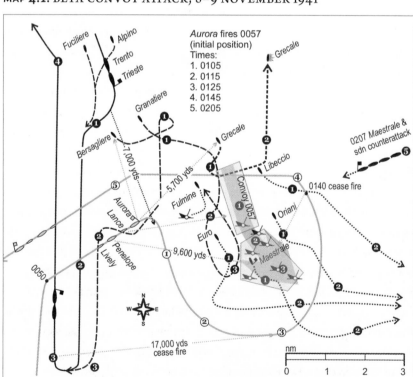

they had been neutralized. He also stressed distributed fire and instructed his captains to refrain from artificial illumination.[44]

These rules were largely defensive, in that they emphasized security and limited the strike force's tactical flexibility, but they certainly worked on 9 November. For all the effort the Italians had put into improving night combat skills, they lacked specific procedures for coordinating the convoy's close escort with the surface support group in the event of a night surface attack, other than to maintain separation. This crippled the Italian defense from the start.

A destroyer from the support group's screen sighted the British from seven thousand meters just before they opened fire and eighteen minutes after the British had sighted the Italians from fourteen thousand yards. HMS *Aurora*, the lead cruiser, laid her guns using radar ranges, but after that used radar only for navigation. The British column first targeted two of the escorting destroyers (fine off the starboard and port bow). These vessels held fire and charged the British column. One was sunk and the other disabled before they reached prescribed torpedo range. The British then turned toward the head of the convoy. The convoy maintained formation, according to instructions, and sprayed flak overhead, as if under aerial attack, even as one freighter after another burst into flames. A third Italian destroyer reached a suitable launch point, but held fire because her captain believed he would be attacking his own cruisers, which bore in the direction from which the gunfire was coming. The Italian escort commander, Captain Ugo Bisciani, on the lead destroyer, *Maestrale*, failed to establish the direction and nature of the attack for precious minutes. "From *Maestrale* we saw two flashes . . . no one could see the trajectories; splashes and ships were not visible. We could not even be sure these were enemy shots and not those of the 3rd Division."[45]

By this time the British were circling around the front of the convoy, and *Maestrale*'s radio antenna had been shot away. Bisciani withdrew to the east, taking the close escort's three surviving destroyers with him. Bisciani wanted to avoid the mistakes of earlier battles and to organize a coordinated counterattack. But this was an unpracticed evolution, and Bisciani was further hampered by lack of a radio. By the time he counterattacked, the British had completely circled the convoy and were returning to Malta. Not one of the seven escorted vessels survived.

And what about the 3rd Cruiser Division? Brivonesi's immediate reaction to the gunfire that erupted off his port bow, not six thousand yards away, was to turn to starboard to gain separation and to "seek clarity."[46] He increased speed from twelve to eighteen knots over a period of ten minutes and engaged at 0103, six minutes after his enemy had opened fire—an eternity in a night action. "Accuracy of aim was difficult because the English ships steaming southward in line ahead, were 'end-on' to the cruiser's gunfire; also, the targets quickly disappeared behind the smoke of the burning steamers and their destroyers' smoke screen."[47] After a few minutes, Brivonesi, desperate to grasp the situation, ordered star shells. Not surprisingly, these fell short and confused the situation rather than revealing the enemy ships. When the cruisers ceased fire at 0125 the range was 17,000 yards, they had expended 207 main battery rounds and 82 shells from their secondary battery but hit nothing. The British never realized that heavy cruisers had been chasing them.[48] Afterward, Brivonesi was court-martialed and found guilty of mistakes, but no "elements of criminal fault." Iachino, in any case, was sympathetic, and even argued to his boss, Admiral Arturo Riccardi, that "you cannot be surprised that the results obtained by inadequately trained destroyers is below expectations."[49]

This action, and a series of smaller successes by Force K, confronted the Regia Marina with one of its great crises of the war. There was no way to avoid a night passage on the way to Africa and—so it seemed—no way to defend a convoy once it had been found by enemy surface forces. In response, Supermarina sailed the entire fleet, including battleships, to protect an important mid-December convoy. It implemented this tactic with great reluctance because oil reserves were being rapidly depleted.[50] Moreover, Supermarina considered convoy escort a misuse of the battleships because it compromised their availability to fight a fleet action. Despite delays and losses caused by collisions and a submarine torpedo in battleship *Vittorio Veneto*, the convoy got through. This general success included an action in which a smaller British force brushed against the vastly superior Italian fleet at dusk. The Italians, from Admiral Iachino to the rawest recruit, believed that the British had declined a night action against a battleship squadron that was alert to their nearby presence. (In fact, the British searched for the convoy, but failed to find it).

Accordingly, *Duilio*, the best of the reconstructed battleships, accompanied the 1942 M43 (3–6 January), T18 (22–25 January), and K7 (21–24 February) convoys. The purpose was "to face in terms of superiority any attack by the enemy light forces."[51] The problem with this solution, however, was fuel. On 1 January 1942 Italian navy fuel reserves stood at 108,000 tons. Additions to the stockpile had averaged 32,000 tons per month over the most recent quarter, while consumption had averaged 81,000 tons a month.[52] *Duilio* consumed three thousand tons of oil just to escort one convoy. Given Italy's increasingly hand-to-mouth fuel situation, battleship escorts were not a permanent solution to the problem of defending convoys at night. British surface forces never attacked a battleship convoy, so how *Duilio* would have fared in a night melee cannot be said.

In all events, during this period the Italian convoys got through. In the last quarter of 1941, when they were under the duress of an effective surface strike force, the Italians delivered 64 percent of the tonnage shipped; during the period of the battleship convoys, the percentage delivered was 94 percent.[53] Many factors were at play. Judicious use of Italian surface strike forces and Axis air forces in the offensive role against the 12 February and 23 March 1942 Malta convoys brought the Regia Marina to the height of its power and influence. This was confirmed by the high percentage of supplies delivered throughout the period and by the results of the British Harpoon/Vigorous Malta convoy operations in mid-June 1942. These saw just two Allied merchant ships arrive out of seventeen that sailed. As will be seen, the greatest Italo-German offensive night surface action success followed during the passage of the Pedestal Convoy though the Sicilian Narrow in mid-August. Meanwhile, there were technical developments.

By spring 1942 the Italian warships could intercept enemy metric radar transmission as a series of beeps using their VHF radio sets, if the devices were tuned correctly. A usable radar set finally had reached the fleet. At the same time, the first of the 62-ton *motosiluranti* (MS) type large motor-torpedo boats entered service. The ability to detect enemy radar and the introduction of their own radar sets led Supermarina to adopt new tactics for offensive night operations, and in June 1942 it planned to send a destroyer flotilla, led by the DeTe-equipped *Legionario*, to intercept the Vigorous convoy at night, if opportunity allowed. No more rake formations and futile solitary chargers; now there would be a column seeking to launch a coordinated salvo of torpedoes, just like the tactics used by German, British, and Japanese destroyers. On 12 August 1942, the German destroyer *ZG3* joined an Italian cruiser-destroyer strike force based at Navarino to intervene in case the Pedestal operation, which was then under way, included a convoy from Alexandria to Malta (which it did not). The German captain was pleased to discover that with a simple exchange of light signals the Italian destroyers could correctly apply his familiar tactics in joint night exercises.[54] None of the Axis ships had radar, but the new German tactics relied on passive sensors, such as the VHF radio and the newly introduced Metox metric detection devices to serve as a nocturnal guide. Supermarina canceled this sortie when it concluded that the eastern convoy was a feint.

To the east, six of the new MS boats, operating in three groups of two, thirteen MAS boats in three groups, and four German S-boats intercepted the Pedestal convoy at night in the Sicilian Straits. It was the most successful motor-torpedo boat engagement of the war on either side—partially because it formed part of a layered attack that had started the day before and had left the convoy damaged and disorganized.

At 0001 on the 13th the convoy's three leading destroyers, paravanes streaming, had just rounded Cape Bon and were headed south. The light cruisers HMS *Manchester* and HMS *Kenya* and three merchant ships followed closely, and another five merchant ships, attended by a single destroyer, were scattered astern. Behind them traveled the damaged tanker *Ohio* and a Hunt-class destroyer in one group and, even farther back, a merchant ship and another Hunt. Strung out along the convoy's route were five damaged ships, including two carriers and two cruisers. Six ships, including an aircraft carrier and a cruiser, already had been sunk.

Four MAS boats attacked the minesweeping destroyers at 0400 and launched two torpedoes before gunfire and searchlights drove them off. The German *S51* operating independently some miles northeast of Cape Bon attacked a tanker and claimed a sinking, but probably missed *Ohio*. Then a pair of motosiluranti tucked up behind the wreck of a British destroyer that had run aground months before, ambushed the large light cruiser *Manchester* at 0104 and hit her with one of their torpedoes. The cruiser sank two hours later. She was the largest warship sunk by an MTB in World War II. As the convoy's remnants continued toward Malta through the remaining hours of darkness, the toll mounted. *MS 31* sank a merchant ship with a pair of torpedoes at 0150. At 0310 and 0340 *MAS 552* and *554* hit *Wairangi*. *Rochester Castle* and *Almeria Lykes* also fell victim to MTBs at about this time. The Germans claimed both, but *Rochester Castle*, which survived, was likely hit by *MAS 564*. *MAS 557* and *553* finished the night by sinking *Santa Elisa*. There were also unsuccessful attacks by *MS 23* and *25* (their commanders were later judged to have been insufficiently aggressive) and *MS 26*.

This battle had several features that distinguished it from previous MTB attacks in the Straits. First, it was part of a larger action that included air and submarine strikes. These caused attrition and disruption that the MTBs exploited. The attackers also benefited from perfect conditions, with a dark, murky, yet calm night. Next, minefields channelized the British advance, reducing the area that needed to be patrolled and permitting the Italians to assign patrol areas along the track allowing for sequential interceptions. In its post-battle assessment Maristat regarded the results as confirmation that the basic theory behind Italy's offensive night combat doctrine was sound, but, as events showed, implementation of that doctrine required practice, adjustments under combat conditions, and a better platform. The fact that it took two years for everything to come together was hardly optimal, but from the Regia Marina's point of view it was better late than never.

Axis domination of the central Mediterranean lasted until the 8 November 1942 Anglo-American invasion of French North Africa. This momentous event, and the subsequent Axis occupation of Tunisian harbors, initiated a new cycle of night actions created by the Axis need to supply another army in Africa. In three weeks, the Italian navy pushed forty-seven ships in thirty convoys to Tunisia with little hindrance. Submarines were ineffective in the narrow and heavily mined waters of the Sicilian Straits, and Cunningham, the British admiral in charge, did not request air strikes against this traffic until 19 November. The siege of Malta had been broken by the arrival of the Stoneage Convoy on 20 November, but not until 27 November did the British base a new Force K at the island. Force Q, three light cruisers and two destroyers, started operating out of Bône in eastern Algeria on 30 November.

On 2 December 1942 Force Q intercepted an important four-ship convoy from Palermo to Bizerte, escorted by three destroyers and two torpedo boats. It was the Beta Convoy all over again, with *Aurora* leading the attacking column around the Axis force and sinking every merchant ship and roughing up the escort. The Italians had known that an enemy strike force was hunting them, which raises the question of how was this possible?

This action was the Mediterranean surface combat debut of British shipborne centimetric radar, and it enabled the British to improve its management of a more dispersed and complicated battle zone. Italian escorts were still charging enemy contacts and holding fire until they came under fire themselves. It was a positive for the Italians that early in the action two of the destroyers put all twelve of their torpedoes into the water at hittable ranges within a span of seven minutes. This was a riposte worthy of better results. Indeed, the destroyers each claimed a hit and until well into the 1950s Italian historians refused to believe that all torpedoes missed.[55] Nine months later, in its evaluation of the action, Maristat noted that "In all actions our units have not used, or only briefly and belatedly used their guns. This shortcoming can be found in all the nocturnal actions fought by our units. . . . Command must be convinced of the great effectiveness of night gunnery in offensive actions (the enemy has repeatedly demonstrated it to us)." The naval staff concluded that the commander should have opened gunfire immediately.[56]

In any case, the battle initiated the campaign of the Route of Death. During this almost six-month period Supermarina applied a number of solutions to the surface threat. First, extensive new minefields neutralized Force Q. While frantic work was under way to develop a new centimetric detector codenamed *Marmotta* (Marmot), the Regia Marina started taking delivery of wartime production escorts *torpediniere di scorta* and *corvette* of the *Ciclone* and *Gabbiano* classes. The Italians also started giving valuable convoys a vanguard force. This system worked well against MTBs and showed utility against surface forces as well when a pair of destroyers, *Pakenham* and *Paladin* (1,640 tons, 4×4-inch, 8×21-inch TT, 36-knots) out of Malta intercepted a small convoy off Trapani on 16 April 1943.

In this action the vanguard torpedo boats, *Cigno* and *Cassiopea* (610 tons, 3×100-mm, 4×450-mm TT, 34 knots) sighted the enemy first from less than nine thousand yards; both sides closed until the Italian leader opened gunfire from twenty-five hundred yards. In an intense twenty-seven-minute gunnery action, the British sank *Cigno* and heavily damaged *Cassiopea*. However, *Cassiopea* hit *Pakenham* hard. Commander Basil Jones led the British force from *Pakenham* and later wrote: "There was no question that the Italians came for us; just as we went for them; and the fire from their second in the line was as vigorous as our own."[57] Hit badly in the engine and boiler rooms, *Pakenham* could not clear the area before daylight and eventually was scuttled. *Paladin* was lightly damaged. The one-ship convoy and its close escort of two torpedo boats escaped unharmed. Costly as it was, this was the most favorable result obtained by the Regia Marina in defending against a night surface attack during the war. It was a matter of aggression and better shooting after all.

During the Sicilian campaign that followed, the nocturnal waters around the large island were contested by fast coastal forces. There were twenty-four brisk night encounters, sometimes at hand-grenade range, involving Italian, British, German, and American MTBs, with few results on either side. The motosiluranti used their Marmotta detectors to initiate action nine times and to avoid combat at least twice. But the true masters of the Sicilian night were the U.S. Navy destroyers, which had four night encounters with the Axis MTBs (on 20–21 July, 3–4 August and twice the next night) and always were able—by radar and training—to dominate the encounter.

CONCLUSION

Italy's night-combat tools, weapons, training, and results were broadly inferior to those of her opponents throughout the war. The systemic reasons for this disparity began with the navy's focus on avoiding night combat except as a means of blockading the Sicilian Straits with coastal and light flotilla units. This meant that Italy's tools, its optics, its night fire-control, and its star shells were geared for that limited use. The Regia Marina did not adequately prepare for the type of night battles in defense of convoys that it was obliged to fight. Once the navy appreciated the extent of its deficiencies, it strenuously tried to catch up, but its enemies were improving even faster, supported by more resources, more experience, better tools, and better cooperation among allies.

Italy's offensive night-fighting platforms, the torpedo and MAS boats, proved inadequate for the jobs that they were given. Torpedoes were used in penny packets. The Japanese fired more torpedoes in surface combat at the Battle of the Java Sea in March 1942 than Italy did during the entire war. Italian night gunnery was too often off the mark despite vigorous efforts to improve it.

In the larger context, Italian lack of success in nocturnal convoy defense cannot be judged in isolation. German convoys along the Norwegian leads, the Breton coast, the Aegean, and the Adriatic certainly fared little better than the Italians when attacked by British surface forces. Between 29 February 1944 and 11 January 1945, British, Canadian, and French forces attacked six German convoys and, suffering nothing worse than machine-gun fire damage to three cruisers, sank eight escorts of at least fleet minesweeper size and severely damaged nine more, as well as six of the nine merchantmen under escort, with two others damaged. Experience in the Pacific and Eastern Indian Ocean between 1942 and 1945 further confirmed that defending a convoy at night was an impossible task *unless* the escort had a technological edge such as radar *and* was much stronger than the attacking force. But even that was no guarantee.[58] Comparing Italian performance to that universe, Italy's record in defending convoys seems average. For Italy the true goal was to manage the overall convoy battle, not to win tactical encounters. The navy could not absolutely protect every convoy against every threat; it had far too much traffic to defend along routes that were subject to repeated attack in the same places. It *was* an impossible job. This finding also affirms that the Italian battleship convoys of December 1941–February 1942 were an appropriate tactic for the situation and forces (if not the fuel levels) then pertaining.

The main thing that can be said about the Regia Marina in this context is that it never gave up, it never gave in, and it remained a factor to be dealt with even at night. The cruise of the new fast-intruder light cruiser, *Scipione Africano*, from La Spezia to Taranto 15–17 July 1943 symbolizes the potential of Italian arms and the quality of the effort that the Regia Marina put into solving many of its technical and night-fighting problems before it ran out of fuel

PHOTO 4.5. *Scipione Africano.* This vessel, with her low silhouette, powerful battery of 135-mm guns, and advanced electronic suite, was the most up-to-date night-fighting platform produced by the Regia Marina in World War II. *Official U.S. Navy photo from Marc'Antonio Bragadin, The Italian Navy in World War II (Annapolis, MD: Naval Institue Press, 1957)*

and time. This vessel—a small, low-silhouette, 39-knot light cruiser, equipped with Gufo radar, centimetric detectors, and rapid-fire 135-mm guns—surprised four British MTBs on the last stage of her journey to Taranto on a calm, moonlit night: ideal MTB conditions. "I was caught completely napping," the British commander wrote, "[W]e never dreamed that a cruiser would be able to get down there unseen through all our patrols." *Scipione* blasted past the British boats, avoiding their torpedoes, and shooting effectively, sinking one and damaging another two. Her captain wrote: "At my order, all of *Scipione*'s weapons, cannons and machine guns, opened fire on them, a fire so precise and violent as to amaze even me, though I knew the ship's capabilities."[59] This was the Regia Marina's last night action against the British involving a vessel larger than an MTB. It was a small victory, but it can serve as a data-point in marking the progress that the Regia Marina made in improving its night combat capabilities.

CHAPTER 5

HOW CAN THEY BE THAT GOOD?
JAPAN, 1922–1942

JONATHAN PARSHALL

The time was just before midnight, 24 August 1927. The place: the Inland Sea of Japan, some twenty miles north of Miho Bay. In the pitch blackness, about thirty Imperial Japanese Navy warships were doing what they seemingly were *always* doing—practicing night fighting. The destroyers of Force B, supported by light cruisers *Naka* and *Jintsū*, were tasked with attacking the heavy units of Force A. Armed with its deadly torpedoes, Force B was charging along at twenty-eight knots, looking for its prey—the battleships *Nagato*, *Mutsu*, *Hyūga*, and *Ise*. However, the exercise umpires suddenly announced that the light cruisers had "attracted enemy searchlights." Misjudging their position relative to their smaller charges, cruisers *Naka* and *Jintsū* quickly reversed course. On board the destroyer *Warabi*, the bridge watch barely had time to shout before the black bulk of *Jintsū* came looming up out of the darkness and slammed into *Warabi*'s starboard side. *Warabi* was sliced in half, burst into flames, and sank immediately with all but a handful of her crew. Meanwhile, *Naka* smashed into *Warabi*'s sister ship, *Ashi*, amputating her stern and drowning another twenty-seven sailors. *Jintsū*'s captain committed suicide shortly before his court-martial.[1] At a eulogy for the lost sailors of the "Mihogaseki Incident," Admiral Katō Kanji, the commander in chief of Combined Fleet, expressed regret for the deaths. But he declared that the strenuous training, "into which the navy has poured its life's blood," would continue, in order to provide "certain victory in the [unequal] struggle."[2] Such was the price that had to be paid to secure the night-fighting edge that the Imperial Navy was certain it needed in any future war against the United States.

STRATEGIC CONTEXT

Japan's prowess in night warfare stemmed from a keenly felt position of weakness. Throughout the interwar period, the Imperial Japanese Navy labored under the certain knowledge that in any confrontation with the United States Navy it would be badly outnumbered. Given the vast disparity in the size and economic power between the two combatants, this could hardly help being the case. Furthermore, these disparities were locked in place by a series of naval limitations treaties that codified Japan's inferiority, forcing its navy to accept (with gritted teeth) a fleet size of between 60 percent and 70 percent as large as those of the U.S. and British navies, depending on the type of vessel in question. Ultimately, these treaties would keep Japan's building programs—as well as those of its enemies—artificially in check until Japan formally left the Treaty system, on 1 January 1937. Although the treaties undoubtedly worked to Japan's benefit from a fiscal standpoint, it was a bitter pill to swallow.[3]

The Japanese found the 5:5:3 ratio of capital ships imposed by the Washington Naval Treaty of 1922 particularly onerous. Not only was it a blow to Japan's national prestige, but according to the naval orthodoxy of the time it doomed the Imperial Navy to ruin in any confrontation with the U.S. battle line. When British engineer Frederick Lanchester published his "N-Squared Law," in 1914, he neatly expressed in algebraic form what it meant to be the weaker opponent in a modern naval engagement. According to Lanchester's equation, the ability of long-range naval guns to enable multiple warships to concentrate all their firepower on a single target meant that a theoretical contest of ten battleships versus six would see the weaker fleet completely annihilated for the cost of only two ships from the stronger force.[4]

Of course, in a real war, Japan could count on the fact that the U.S. was unlikely to be able to muster *all* of its capital ships against the Imperial Navy. Likewise, matters of distance and logistics were assumed to weaken an advancing force by 10 percent for every thousand miles traveled. Given this, the U.S. calculated that even after a westward advance, it would still fight with 70 percent of its strength, hence the need for the Japanese to be negotiated down to 60 percent. Not surprisingly, the Imperial Navy, which could do its sums just as well, wished to fight at parity, which it could achieve if its fleet were at 70 percent of America's.[5] Receiving a limit of only 60 percent in the treaty negotiations was thus seen as both humiliating and a direct threat to Japan's security. The Imperial Navy's problem, then, became figuring out how to do more with less.

Ever since the successful conclusion of the Russo-Japanese War in 1905, it was accepted as gospel that the outcome of any future war with America would be decided by a "Decisive Battle"—an enormous fleet-on-fleet action (*Kantai Kessen*) involving a clash between both sides' battleship forces.[6] In other words, Japan would have to fight—and win—another Tsushima-like confrontation. But given the likelihood that it would be outnumbered by a disastrous 5-to-3 ratio in capital ships, how could the odds be evened?

Japan's answer to this problem was two-fold. The first principle was to look to technology for equalizers. In this sense, the Japanese were avowedly pro-technology and pro-innovation. With respect to Japan's battle line, the result was an obsession with being able to overcome quantity with quality—"*Ka o motte shū o sei-su* [Using a Few to Conquer Many]" in Imperial Japanese Navy parlance. This led directly to the Imperial Navy's obsession with outranging the enemy.[7] During the 1930s, the turrets on Japan's battleships were all upgraded to increase their elevation, and thus their firing range. Sophisticated fire-control directors were placed ever higher in the ship's structure so as to enable the ship's guns to direct devastating long-range fire against the enemy more accurately. The logical conclusion of all these drivers was the laying down in 1937 of the enormous *Yamato*-class super-battleships; their huge 46-cm (18.1-in) guns would give the ability to outrange and outclass any American battleship.

Just improving the battle line might not be enough, however. This led Japan to embrace a second principle—a push to widen the parameters of the main fleet-on-fleet engagement away from a strictly daylight encounter to one that included nocturnal battles. By creating a nocturnal attritional phase before the anticipated daylight gunnery action, the Japanese hoped to whittle the American main battle line down to size before the decisive clash. In this new environment, the Japanese felt that the twin principles of quality over quantity and outranging also could be put to good use. With all this in mind, starting in the mid-1920s the Japanese threw themselves determinedly into mastering night combat.

TECHNOLOGY AS DRIVER

Fighting effectively at night required specialized tools, and Japan labored mightily throughout the interwar period to optimize a wide array of technologies to meet its particular needs.

Optics

The first, most fundamental problem in night combat is simply seeing well enough to fight. Before the advent of radar, perhaps the best solution to that dilemma was fielding better night optical equipment. Here, the Japanese utilized their time-honored strategy of first adopting foreign-made technology and then moving to licensed manufacture, all the while nurturing home-grown capabilities that would eventually take over. Beginning around 1915, Imperial Navy contracts were directly responsible for the growth of Nikon, Canon, Minolta, and Fuji Film.[8] It was Nikon that eventually developed a range of superb spotting binoculars in oversized 8-cm, 12-cm, and 21-cm sizes. The increased aperture size of these instruments allowed them to gather as much light as possible. To maximize their potential still further, the Imperial Navy used hand-picked, rigorously trained night lookouts. As a result, Japanese warships had a visual detection capability that often rivaled, and sometimes exceeded, even that of mid-war radar sets.

Illumination

Once a target was detected, gunfire was the first and most obvious means to destroy it. Guns had the advantage of being quickly laid on a target, and could thus adapt to rapidly changing tactical circumstances. On the other hand, gunfire had the downside of being very difficult to aim at night without some sort of illumination to provide adequate light for both target-designation and ranging purposes. To solve this problem, the Japanese relied on both searchlights and star shells.

Searchlights

Early Japanese warships were equipped with German Siemens-Schuckert searchlights. By the end of World War I, though, the Japanese navy was using British Sperry 150-cm units.[9] These lights, capable of generating nine thousand candlepower, theoretically enabled the Japanese to aim their guns to about seven thousand meters, although the practical engagement range was still below two thousand meters.[10] By the time of the Pacific War, though, the Japanese navy's ships all used powerful domestically produced models developing up to 13,600 candlepower. A single light could illuminate targets out to eight thousand meters; two lights with overlapping beams could do the same out to ten thousand meters. It was recognized that searchlights had the drawback of revealing the position of the firing warship. Accordingly, the Imperial Navy developed sophisticated searchlight-control arrangements, enabling a central searchlight director to pre-aim lights onto the correct bearing automatically and then suddenly illuminate the target.[11] Once the victim was smothered with gunfire and dispatched—or was sufficiently illuminated by its own fires—the searchlights would be switched back off to cloak the firing vessel in darkness once more.

Star Shells

Unlike searchlights, star shells had the advantage of being somewhat less revealing of the firing ship's position. The problem was in providing both a sufficiently bright light and having it persist long enough to be of help to gun crews. Beginning in 1921, the Japanese introduced star shells (*seidan*) that burned as they fell. These "A"-type shells produced a great deal of light, but they only lasted a few seconds. Beginning in the late 1920s, though, experiments with parachute shells were carried out, leading to the introduction in 1935 of the "B"-type (Shomeidan Otsu) illumination shells. The largest of these, fired by the *Nagato*-class battleships, were capable of hanging over the battlefield for up to a hundred seconds, while pumping out a dazzling 5.3 million candlepower. Even the shells used by destroyers could generate 680,000 candlepower for 55 seconds. Shells such as these could literally turn night into day. Taken together, better optics and illumination meant that by the eve of the Pacific War Japanese guns could reliably engage targets at night at distances up to eight thousand meters—a four-fold increase in the space of twenty years.[12]

Propellants

Muzzle flashes from the guns also ran the risk of giving away the position of the firing warship. By 1938, the Japanese had partially solved that problem by developing flash-suppressing powder that made the gun's muzzle blast much more difficult to spot. By adding potassium sulfate to cordite, they largely eliminated the flashes in guns of up to 5.5-in caliber, thus shrouding the fire of their destroyers and older light cruisers.[13]

COMMAND AND COMMUNICATIONS

Less is known about Japanese naval communications than about other, more "glamorous" areas of their naval technical development. But the Imperial Navy's capabilities were certainly not lacking. At the operational level, Japan's naval communications network was built on an interconnected hub-and-spoke model, while regional communications centers were responsible for broadcasting orders to—and receiving communications from—any warships at sea in their operational area.[14] Pertinent information from task forces at sea was first sent to the regional hub, which then relayed them up to the 1st Communications Unit in Tokyo. This unit served as the master communications node for all of Combined Fleet, rebroadcasting any particularly pertinent information back out to the regional nodes, which in turn would broadcast them to any subordinate task forces that needed to be in the loop. As the Americans noted after the war, the arrangement resulted in "a complex, modern system, highly flexible and efficient."[15]

At the task force level the Japanese relied on radios, signal lights, and flags. The latter, of course, were far less useful after nightfall, and in practice the only visual communications methods used at night were 60-cm searchlights used for signaling.[16] Japanese warships were well-equipped with radios, however, and the Imperial Navy continually upgraded its equipment throughout the interwar period.[17] By the time of the Pacific War, a typical cruiser flagship carried no fewer than fifty-five radio sets and a pair of radiotelephones.[18] The latter were capable of transmitting on VHF bands, providing secure line-of-sight (i.e., short-range) tactical communications.[19] Although postwar American technical assessors were largely unimpressed with the physical characteristics of much Japanese radio equipment, they did acknowledge that "the newest enemy equipment compares favorably with our models."[20] The Japanese also conducted experiments with infrared signaling equipment starting in 1941, but the gear was not formally introduced until July 1943.[21]

Radar

One crucial area where the Imperial Navy lagged was in the development of radar.[22] Although Japanese scientists were well acquainted with the theoretical fundamentals of radar technology, they did not progress as quickly as those of other major industrial nations. A Japanese technical mission to Germany in early 1941 had revealed the superiority of the Reich's shipborne FuMO sets, which finally spurred the Imperial Navy to kick its own research into high gear.[23] Japan's efforts were hampered, however, by the insistence of its navy and army to maintain completely separate research operations. Nor did the services do a good job of harnessing the expertise of leading civilian researchers.[24] As a result, the Navy's first operational radar equipment—an air-search set on battleship *Ise,* and a surface-search set on her sister ship *Hyūga*—was not sent to sea until May 1942.[25] By this time Japan was at least two to three years behind both the British and Americans.

This late adoption of radar also appears to have delayed the introduction of the shipboard information-processing facilities needed to assimilate the new sensor inputs. The Japanese navy did eventually install an "information room" in some of its cruisers that was analogous to the U.S. Navy's Combat Information Center.[26] However, this seems to have occurred around 1944, at a time when the importance of surface combat in the naval war was receding.[27]

Torpedoes

Unquestionably the area where the Japanese made the most progress was with torpedoes. The Imperial Navy had made good use of torpedo boats during the Sino-Japanese War of 1894–95, launching several spectacular attacks against Chinese warships in harbor.[28] But employing these early model fish effectively required nerves of steel. Torpedo boats would race in to deliver their ordnance from ranges of just a few hundred meters against targets that were often blazing away with everything they had. "Nikuhaku-hitchū! [Press closely, strike home!]" became the watchword of Japan's torpedo boat sailors, who knew it was their likely fate to lead short, exciting lives in combat. Indeed, even up until the 1930s, daylight torpedo attacks were still being delivered from less than five thousand meters, with nighttime attacks being launched from less than two thousand.[29] All of this began changing radically, though, in the 1930s.

The Type 93 Torpedo

Torpedoes are inherently complex, finicky beasts, crammed full of gyroscopes, fuel systems, pumps, cooling systems, warhead fuzes (both contact and later magnetic), and many other intricate mechanisms. The motor is crucial, since it must pack as much mechanical energy into as compact a space as possible. Wringing every extra horsepower out of the engine can enable the fish to go farther, or faster, or carry a larger warhead. Unsurprisingly, having placed large bets on night combat, the Imperial Navy wanted radical improvements across all three dimensions.

The simplest way to make a more powerful weapon was to make it bigger. Accordingly, the Japanese had begun working on a 61-cm (24-in) torpedo just prior to the signing of the Washington Naval Treaty. This Type 8 torpedo was based on conventional British Whitehead models. It was unremarkable technologically, but it carried a large warhead, and had a range a third greater than comparable 53-cm (21-in) weapons.[30] The Japanese navy quietly deployed it on its destroyers and light cruisers, despite the fact that the Washington Treaty forbade torpedoes larger than 53-cm.[31] An improved model, the Type 90, was fielded in 1930. By then, the Imperial Navy possessed torpedoes whose performance "was as good as that of any in the world."[32]

Still, a truly radical improvement in torpedo performance had to start with the power plant. Conventional torpedoes of the day burned a mixture of kerosene fuel and compressed air. Normal air is 77 percent nitrogen, 21 percent oxygen, and the remainder argon and carbon dioxide (CO_2). Only the oxygen is useful in combustion; the other elements simply slough through the combustion chamber and then bubble up to the surface as a wake that gives away the torpedo's presence. But if pure oxygen could be used, enormous benefits in the efficiency of the power plant could be realized.

The Japanese had begun working on pure-oxygen torpedoes as early as 1916, but these first experiments were unsuccessful and the notion was shelved.[33] In 1927–28, though, Japan became concerned that the British apparently were working on large 24.5-in enriched-air torpedoes for their new battleships, HMS *Nelson* and HMS *Rodney*.[34] This spurred a renewed interest in the topic, and a team from the Imperial Navy's Torpedo Experimental Division—including Captains Kishimoto Kaneharu and Yamashita Kanemitsu and Lieutenant Commander Oyagi Shizuo—was assigned the mission of making an oxygen torpedo a reality.[35]

Working with pure oxygen presented a number of formidable problems. First and foremost, oxygen tends to react spontaneously and vigorously (which is a polite way of saying it explodes) if it comes into contact with hydrocarbons such as oil or grease. Unsurprisingly, this is not uncommon in something like a fuel line or a valve. Likewise, even minor machining imperfections in these fittings can leave rough metal edges or burrs. When these are subjected to a rapidly moving, high-pressure gas, they can be heated to the glowing point by friction alone, leading to the same unhappy results. Even sharp turns in the fuel lines create pressure waves causing the oxygen to heat up. In other words, using pure oxygen took the normally finicky nature of torpedo design to a new level of persnickitiness.

The heart of the solution lay in extraordinarily careful manufacturing. The torpedo's fuel lines were redesigned to smooth tight turns, finely machined

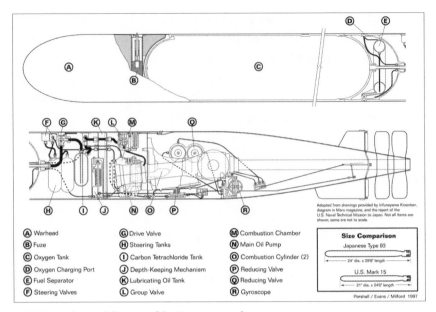

FIGURE 5.1. Internal diagram of the Type 93 torpedo

to eliminate any pits or scratches, and then scrubbed with a potassium compound to remove any hydrocarbons left over from the machining process.[36] In order to eliminate explosions at the time the motor was started, a small tank of normal air was used to initiate the combustion process. Once the motor was running, the starter air was reduced until the engine was using pure oxygen. By 1932, the team had a working motor. A year later, prototypes of the actual weapon had been manufactured for test-firing on board the cruiser *Chōkai*. Further refinements over the course of the next two years resulted in the world's first fully oxygen-powered torpedo: the Type 93 Model 1.

Commonly known after the war as the "Long Lance," the Type 93 was hands down the finest torpedo of World War II. It was huge—9 meters long (29.5 feet), weighing 2,700 kg (5,952 pounds), and carrying a whopping 490-kg (1,080-pound) warhead. Powered by kerosene and 300 kg of pure oxygen compressed to 3,200 psi, the new version effectively had eliminated more than a ton of nitrogen from the torpedo's weight profile.[37] Its engine was extremely powerful—520 brake horsepower—capable of hurling the torpedo through the water at up to 50 knots.[38] It left practically no wake; only the slightest sheen of lubricating oil was visible at short range in daylight under the right conditions.[39] At night, it was utterly invisible. What made the Type 93 such a game-changer, however, was its extraordinary range—up to 40,000 meters, at a time when contemporary 53-cm torpedoes had a maximum range of perhaps eight thousand meters. This exceeded the effective range of the very largest battleship guns, and gave the Japanese navy a long-range underwater strike capability that no other navy possessed.

The Imperial Navy also had succeeded in hiding this leap in performance from its rivals. No one in the U.S. Navy would suspect that the Japanese had such a weapon until mid-war. In the after-action report from the Battle of Kolombangara, written in August 1943, the American task force commander acknowledged the respect that he and his skippers had for Japanese torpedo tactics "and the wallop possessed by their torpedoes." He then mentioned: "I have been informed that a recent Japanese torpedo recovered off Cape Esperance [on Guadalcanal] is a 24-inch torpedo 24-feet long. . . . [T]his would indicate that [the Japanese] may have a torpedo that will make 40 knots for 10,000 yards."[40] Even this guess, though, underestimated the Type 93's highest speed by ten knots, and missed its maximum-speed range by more than half. The Type 93, in other words, was one of the war's few examples of a truly secret weapon.

PHOTO 5.1. The first captured Type 93 torpedo, photographed shortly after its discovery at Cape Esperance on Guadalcanal. Up until 1945, this was the sole example of the weapon recovered by the Americans. *U.S. Naval History and Heritage Command*

Torpedo Reloading Systems

It was recognized that in the heat of battle more than one salvo of torpedoes might be needed, and that warships should carry reloads. At the same time, though, just *moving* one of these three-ton monsters from a storage rack to the tubes—at night, under fire, and on a pitching, rolling, radically maneuvering warship—presented a significant challenge. By the mid-1930s, however, the Japanese navy had perfected reloading gear that enabled crews to extract its 61-cm fish from a storage box located adjacent to the mount. Secured on rollers, the torpedo could then be winched into the tube, allowing the entire mount to be reloaded in as little as three minutes.[41] In the heavy cruisers, overhead tracks and pulleys permitted reloads to be moved quickly from storage to any of the ship's mounts.[42]

Torpedo Fire-Control

As Japanese torpedo ranges increased throughout the 1930s, the Imperial Navy also upgraded its fire-control systems.[43] By the time the Type 93 arrived, the Japanese navy was utilizing torpedo directors every bit as sophisticated as anything found in the West, and capable of blind-fire if the target became shrouded from view by a smoke screen or cloud bank. Postwar American observers were impressed by the entire system, calling it "surprisingly complete and comprehensive, provision having been made for introducing into the computing unit almost every conceivable constant and variable pertaining to the problem." Indeed, in some ways, Japanese torpedo fire-control systems were more sophisticated than their gunfire systems, in that they used automatic inputs combined with electrical follow-up gear.[44]

In sum, the arrival of the Type 93 torpedo, sophisticated long-range fire-control systems, and rapid salvo firing offered possibilities that simply had not been available. Even before the Type 93 was developed, though, Japanese warship designs had been optimized for torpedo warfare.

WARSHIP DESIGN

In its purest form, the design of a warship should represent the physical embodiment of its navy's doctrine. In this sense, the Imperial Navy produced warships that were remarkably coherent expressions of its doctrinal preferences—speed and firepower (particularly torpedo firepower) at the expense of habitability, protection, and survivability. They were practically odes to the primacy of the offense. With respect to night combat, the two most important Japanese ship types were destroyers and heavy cruisers.[45]

Destroyers

Recognizing the primacy of the destroyer to its torpedo tactics, and in keeping with their philosophy of using a few to vanquish many, the Imperial Navy decided in the mid-1920s to significantly up the ante. The result of this drive for technological dominance was the *Fubuki*-class destroyer.[46] When *Fubuki* was launched, in November 1927, she was in some ways as significant a leap forward as the HMS *Dreadnought* had been to battleships twenty years before. *Fubuki* displaced about 50 percent more than a World War I–era destroyer. Her enormous powerplant—50,000 shaft horsepower—was twice as powerful as most contemporaries, driving her along at thirty-eight knots. *Fubuki's* gun power was unprecedented—six powerful 127-mm/50-cal. guns that could (theoretically, at least) also engage aircraft.[47] The guns were housed in enclosed twin turrets that were waterproof and gastight and offered splinter protection. Unlike the open mounts of her predecessors, which relied on manual loading, *Fubuki's* turrets were fed by ammunition hoists from her magazines, increasing her rate of fire significantly.

Fubuki's true claim to fame, though, was her heavy torpedo armament— nine 61-cm tubes in three triple mounts. Another nine reloads were carried. Even if she had mounted 53-cm tubes, *Fubuki* would have had the heaviest torpedo broadside of any warship afloat. The fact that she was secretly armed

PHOTO 5.2. A fine study of the *Fubuki*-class destroyer *Shikinami* at speed. Her turrets (one forward, two aft) and torpedo tubes (one between the funnels, two between the aft funnel and "X" turret) are all clearly visible. Note, too, the long forecastle and flared bow for better seakeeping and the distinctive air intakes around both funnels. *Fukui Collection*

with powerful 61-cm Type 90 weapons simply made the gap that much more pronounced. Put simply, the Japanese had just launched a destroyer that carried more raw firepower than most countries' light cruisers. Indeed, *Fubuki* represented such a quantum leap that the Japanese coined an entirely new moniker for her: "*Toku gata*," meaning "special-type" destroyer. She was to be the prototype for every Japanese destroyer commissioned until 1942.

Unsurprisingly, the emergence of these formidable vessels sent ripples of consternation through the world's navies—particularly those of the British and the Americans. As is the nature of most military technologies, it was not long before Japan's rivals began laying down larger, more capable destroyers of their own. Indeed, at one point the Japanese briefly stopped building more *Toku gatas*, in order to head off the naval arms race they had just initiated. But it was too late for that—the die had been cast.

What foreign observers were unaware of were *Fubuki*'s limitations. In order to cram the necessary powerplant and armament into a compact hull, Japanese designers had employed every possible weight-saving measure. *Fubuki*'s machinery was lighter than originally specified, and her speed was lower. Her upper works were built with lighter, weaker alloys, and her hull made extensive use of welding (which was still an emerging technology for naval construction) and butt joints, rather than standard overlapping riveted joints. This improved the ships' weight and speed, but unknown to *Fubuki*'s builders it also diminished her structural strength. Even with this, *Fubuki* still was two hundred tons overweight when she was launched—much of it topside, which lowered her stability.[48]

Fubuki was also not a well-rounded design. Where most navies' destroyers were jacks-of-all-trades that could perform surface attack, defensive screening, and antisubmarine work, *Fubuki* was fundamentally skewed toward offense.[49] Her sonar was rudimentary, she carried only a small number of depth charges, and her main armament eventually would be shown to be inadequate against aircraft.[50] She was a magnificent surface-warfare platform, but that was it. However, these deficiencies would not become apparent until later.

Heavy Cruisers

With the Washington Naval Treaty's battleship moratorium, the focus of the naval armaments race had switched to the 10,000-ton "Treaty" cruiser designs. Even before the Washington Treaty was signed, the Japanese had already launched their small 3,000-ton *Yubari*, and the first two 7,500-ton *Furutakas*. These had proven useful in prototyping the hull shape and other construction characteristics of Japan's first true Treaty cruisers—the *Myōkō* class. The *Myōkōs*, in turn, established the archetype for every Japanese cruiser built up to the outbreak of the Pacific War—big, fast, armed to the teeth, and badly overloaded.

Myōkō carried ten 20-cm guns in five twin turrets, thereby upping the ante on British and American heavy cruisers, which typically carried eight or nine such weapons.[51] Her secondary armament of six 120-mm dual-purpose (DP) guns was later upgraded to eight capable 127-mm DP weapons. It was her torpedo battery, however, that completely eclipsed that of her rivals. Originally sporting twelve fixed 61-cm tubes, she was eventually upgraded to four quadruple 61-cm mounts, with eight reloads.[52]

There actually had been vigorous debate between the Naval Construction Section and the Naval General Staff regarding the wisdom of carrying any torpedoes whatsoever.[53] The presence of large torpedo warheads—located directly amidships, above the main armored deck, and without any real protection—posed a fearful risk in the eyes of Japan's chief naval constructor, Captain Hiraga Yuzuru. However, the tactical imperatives of the insistent General Staff won out, and the torpedoes were fitted.

Eventually, *Myōkō*'s upgraded torpedo tubes would be moved slightly outboard on sponsons, thereby theoretically reducing their danger to the ship.[54] Torpedo reloads were also eventually stored in armored boxes using 25-mm high-tensile Ducol steel to shield them from fragments.[55] In some later "A-class" cruisers, the torpedo warheads were stored separately beneath the armored deck, and then only joined to the torpedo body prior to battle.[56] The "A-class" cruisers, too, subdivided the torpedo room containing the ship's mounts, using fireproof curtains. Yet, all of these defensive measures proved little more than cosmetics. As war experience would demonstrate, Hiraga's fears of torpedoes exploding as a result of battle damage were well-justified.

Myōkō's hull featured two unique Japanese innovations. The first was forgoing backing plates for her armor and instead integrating it directly into the hull as longitudinal strength elements, thereby saving weight. The distinctive undulating line of her weather deck, with a pronounced sheer and flare forward for seakeeping, a straight line amidships, and then a pronounced downward

PHOTO 5.3. The superstructure of heavy cruiser *Takao* in 1939. Its enormous size—necessitated by command and staff accommodations befitting a squadron flagship—is clearly evident. The superstructure is festooned with a variety of rangefinders, fire-control directors, and searchlight command stations. One of the cruiser's Type 92 quadruple 61-cm torpedo mounts is trained to starboard, showing how they overhung the ship's hull when in firing position. The danger to the ship's own command spaces from an induced torpedo detonation is self-evident. *Fukui Collection*

rake toward the stern, likewise increased strength and saved weight.[57] Despite these measures, however, *Myōkō* came in a whopping 12 percent overweight. Although the Japanese were not overly concerned about strictly toeing the line when it came to the Treaty's tonnage restrictions, such a gross discrepancy cannot have been intentional, since it lowered *Myōkō's* freeboard, reserve buoyancy, and speed, and also negatively affected the ship's stability.[58] Much of that weight was topside. Her large superstructure, for instance, was driven by the need to concentrate the ship's command, communication, and numerous fire-control instruments in a single location. Indeed, *Myōkō's* successor, the *Takao*-class cruisers, would have even larger, more elaborate superstructures.[59] To foreign observers, though, unaware of the design's problems, the picture *Myōkō* presented was formidable. And despite their problems, the Imperial Navy's heavy cruisers would demonstrate during the war that they were extremely tough customers, capable of dishing out and absorbing a great deal of punishment.

DOCTRINE

Throughout this time period, as both technology and warship design were interacting, Japanese thinking on the matter of night combat had been evolving as well. By 1930, the first Toku gata destroyers had been launched, with their heavy torpedo armament. The Japanese now settled on the destroyer squadron—roughly twelve to sixteen ships—as their basic attack unit. These would be led by a 5,500-ton light cruiser, which had the communications facilities and extra accommodations to serve as a flotilla flagship.[60] Given their heavy armament, a single squadron of "Special Type" destroyers was capable of firing a salvo of up to 144 torpedoes. Accordingly, Japanese destroyer tactics began focusing not just on attacking their opposite numbers, but on penetrating the enemy screen and attacking his main battle force as well.

Even as this evolution was taking place, the Japanese had been considering the proper role of their heavy cruisers. That role, oddly enough, turned out to be support. In order for the destroyers to do their business against the enemy's battle line, it would be necessary to destroy the enemy screen. But that screen likely would comprise enemy cruisers as well as destroyers, with the cruisers having enough firepower to defeat or at least seriously degrade the Toku gata squadrons before they could deliver their decisive torpedo attacks. As such, Japan's new heavy cruisers were seen as the means of overwhelming the enemy screen. Likewise, their large command and control capacity also would help support the destroyer squadrons; indeed, they could act as flagships themselves and help coordinate larger multisquadron battles.[61]

The period between 1934 and 1936 saw a rapid evolution of the fleet's doctrine. All parties within the navy agreed that the introduction of the stupendous Type 93 torpedo clearly necessitated a rethinking of how the Decisive Battle should unfold. But there was still wide-ranging discussion about where nighttime combat fit in the wider scheme of that Decisive Battle, and what role the navy's capital ships might play after the sun went down.[62] By 1936, though, the critical importance of night combat had been acknowledged. The Decisive Battle was now seen as at least a twenty-four-hour affair: spotting the enemy during the day, then preliminary twilight skirmishing, followed by an all-out, coordinated night attack against the enemy's main body. The following morning, with the enemy then badly disorganized and attrited, the Japanese navy's heavy battleships would engage and finish the job.

In order to support the night phase, an entirely new unit was created—the Night Battle Force (*Yasen Butai*)—corresponding roughly with the navy's Second Fleet. It was to comprise all of the torpedo attack units (including several torpedo flotillas), all of the heavy cruisers, and (remarkably) the recently reconstructed fast battleships of Battleship Division (BatDiv) 4 (*Kongō* and *Kirishima* initially, later also *Haruna* and *Hiei*).[63] Originally designed as World War I–era battlecruisers, the *Kongōs* no longer were well-suited to slugging it out against more heavily armored American battleships. However, at night, and against enemy cruisers and destroyers, their handicaps would be minimized. Swift, powerful, and with excellent command and control facilities, their heavy gun power would help blast a way through the enemy screen. Thus, 1936 saw the Japanese arriving at a doctrinal outlook that was the inverse of what might be thought of as the standard orthodoxy, in that it envisioned the usage of *capital ships* to help support *destroyers* in the fulfillment of the latter's attack mission.[64]

Tactics

In order to execute what effectively was becoming a Decisive Battle in itself, the Japanese evolved a series of tactical evolutions to implement the 1936 night-battle concept. An example of one such plan envisaged the enemy fleet being sighted in the afternoon, whereupon the Japanese units would reconfigure from dispersed scouting lines into their concentrated night-attack formations.[65] The night-battle commander would take control, ordering the Yasen Butai to begin stealthily enveloping the enemy fleet as night came down. Once correctly positioned, the attack flotillas would be ordered to commence simultaneous long-range concealed torpedo fire, laying down an intersecting weave of more than a hundred Type 93s. The enemy would have no indication of the stealthy underwater assault that had just been launched. The attack units would reload tubes. A trailing formation would head off the enemy should it try to escape.

Just before the first shoal of torpedoes was timed to hit, the night-battle commander would order an all-out attack. As the initial torpedo salvo struck home, the Yasen Butai would close in from all sides, guided by star shells and night-scouting aircraft. Assuming a 15 percent hit rate, between fifteen and twenty enemy warships would be damaged or destroyed by the initial barrage. BatDiv 4's heavy guns, along with the heavy cruisers, would then blast holes

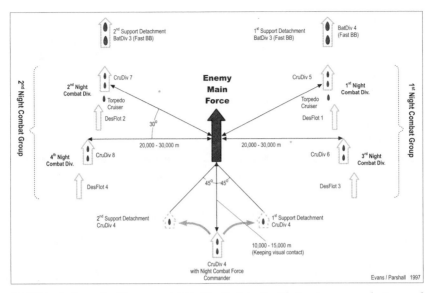

FIGURE 5.2. Example of Japanese night-battle attack tactics. Their extreme complexity—and unreality—is apparent.

through the American screen, and the destroyer flotillas would race into the heart of the enemy formation. This final assault was to be delivered very much in the spirit of *"Nikuhaku-hitchū!* [Press closely, strike home!]," with fish being launched from as little as two thousand meters. The result would be the destruction of a good percentage of the enemy force, along with their cohesion and morale. Indeed, if things went well, the Japanese main battleship force might be committed now as well.[66] Most likely, though, the final drama would unfold the following morning. As dawn loomed, the surviving Japanese ships would re-form their day battle formations, the big battleships of First Fleet would move up to take on their American opposite numbers, and the enemy fleet would be annihilated.

Unfortunately for the Imperial Navy, there were a number of problems with this scheme for nautical *Götterdämmerung.* The first was its incredible complexity. The blithe expectation that a dozen different night-battle squadrons would somehow manage to deploy themselves smoothly around the enemy's perimeter—in the dark, at high speed, possibly in foul weather, without the benefit of radar or active sensors of any kind, and without tipping off the enemy—was (to put it mildly) rather optimistic. The second was the incredibly passive nature of the enemy, which was presumed to go meekly to its destruction without getting wind of what was afoot, or without reacting to it in apparently any meaningful way. Taken together, it is right to view such a tightly choreographed plan's chances of success in actual combat with a healthy dose of skepticism.

This points to a larger issue: by the outbreak of the Pacific War, the Japanese navy had been badly warped by its own battle doctrine. Bit by bit, the crushing tectonic pressure of having to face up to America's overwhelming economic power had suborned almost every unit of the Japanese fleet into playing some role—and *always an attack role*—in the Decisive Battle. Battleships and heavy cruisers were now supposed to sacrifice themselves in support of destroyer attack squadrons. They, in turn, were expected to accept their own heavy losses as they pressed home their assaults. While all this was going on, Japan's fleet submarines would be targeting enemy capital ships. Submarine tenders would be releasing schools of minisubs into the battle as well. By the eve of war, the Japanese had even modified a pair of light cruisers to carry forty 61-cm tubes for use as torpedo cruisers. Everything—*everything*—depended on the outcome of this single gigantic confrontation.

But in service of this battle, the Imperial Navy had neglected other classic roles that great navies have always fulfilled, including commerce destruction, commerce protection, and amphibious force projection, to name a few. More important, the Imperial Navy had not stopped to consider adequately whether the nature of the *next* war would actually be congruent with the *previous* war. Put simply, if the United States was not the same sort of animal as czarist Russia—that is, if it could not actually be knocked out of the war as the result of a massive defeat—things were not likely to end well.

This is not to say that the Imperial Navy was not a formidable opponent; it most certainly was. At the tactical level, by the eve of the Pacific War the Japanese navy had few equals; in the arena of night combat, it had none. But weapons and tactics alone do not win wars, as the Japanese would discover to their cost.

Training

In order to carry out its doctrine, the Imperial Navy trained its men relentlessly. Japanese sailors joked that their weeks contained two Fridays, two Mondays, but no weekend.[67] Ships exercised under frighteningly realistic conditions, at night, and in foul weather. If the Mihogaseki Incident was exceptional, it was also emblematic of how seriously the Japanese navy took its profession. A sailor being swept overboard during an exercise barely rated comment; collisions and hair-raising near misses were just the cost of doing business. Sure enough, just a few years after the disaster at Mihogaseki, a collision between the *Fubuki*-class destroyers *Inazuma* and *Miyuki* on 29 June 1934 led to the latter's sinking—fortunately this time with less loss of life.[68]

However, it was another pair of exercises that not only highlighted the dangers of realistic training, but pointed to systemic structural issues in the Navy's warships. On 12 March 1934, the new torpedo boat *Tomozuru* capsized in a gale during an exercise. Discovered the next morning still floating bottom up, she was towed into Sasebo. Only thirteen men were brought out of her hull alive; another hundred drowned. *Tomozuru*'s basic design philosophy— cramming about half of *Fubuki*'s armament into a third of her displacement— was already suspected of having created a top-heavy design.[69] Now, perceptive observers inside the navy began wondering if the relentless pressure from the Naval General Staff had not created more widespread problems.

Then, in September 1935, during a major exercise in the North Pacific, a large group of warships from Japan's Fourth Fleet ran smack into a typhoon. At the storm's height, with seventy-plus-knot winds and eighteen-meter seas, many ships began rolling alarmingly. Some of the *Fubuki*-class destroyers heeled as much as 75 degrees. Two *Fubuki*s lost their bows at precisely the same weld line; many others suffered structural damage. The hull of the brand-new heavy cruiser *Mogami* was distorted. Joints on her sister ship, *Mikuma*, the four *Myōkō*s, and the tender *Taigei* were all loosened as well. Fifty-four men were lost, and several ships had to be towed to port.[70]

There could no longer by any doubt: the Japanese navy's warships were structurally unsound. Too many years of the Naval Construction Section kowtowing to the General Staff's demands had led to disaster. A crash program was undertaken to rebuild many of the navy's ships.[71] Superstructures were cut down and weapons redistributed or removed to reduce topweight; stiffeners were added to strengthen hulls; welding guidelines were revised, and in some cases entire hull sections were re-riveted; bulges and ballast were added down low to improve stability. The final result was that the Japanese warships emerged on the eve of war in far better shape structurally. But its building programs were probably set back by at least two years. By the time the repair work was completed, most of the warships that would fight in the crucial early period of World War II had already been either built or ordered.

EARLY WAR NIGHT BATTLES

The year 1942 was the time when the Imperial Navy's combination of excellent training, flexible tactics, and superior night-battle technology came together to create several notable victories. Japan's prowess at night combat came as a stunning surprise to the Allied navies, which had not imagined they were nearly so capable. Yet, even with all the tools in its night-fighting kit, the Japanese navy built a track record that was far from perfect as its strengths and weaknesses began interacting with those of its chief rival—the U.S. Navy. What follows is not a complete description of these battles, but rather their more interesting facets from the standpoint of night combat.

The Battle of the Java Sea

The invasion of Java and Bali represented the climax of Japan's first-phase military operations. On 27 February, as a Japanese invasion convoy began moving toward western Java, the Allies gathered their few remaining naval assets.[72] A squadron of two heavy and three light cruisers and nine destroyers under Dutch Rear Admiral Karel Doorman sortied north from Surabaya in a desperate attempt to locate and destroy the Japanese convoy. Rear Admiral Takagi Takeo and his pair of heavy cruisers showed up just before Doorman could intercept it, joining the two Japanese light cruisers and fourteen destroyers already screening the convoy. The Allies were stronger in gun power, but Takagi's warships were faster, allowing him to dictate the range. They also carried 280 powerful 61-cm torpedoes—more than enough to destroy Doorman's force if they performed as the Japanese expected.[73] From a doctrinal standpoint, this was a very close match to Takagi's prewar training—a classic stand-up fight between two small fleets, each with a cruiser main body of sorts and their screening elements. Even so, the battle itself devolved into a confusing eight-hour affair running from 1600 to almost midnight.

The initial action saw Takagi opening with very long-range gunfire (from 23,000 meters to 26,000 meters) against Doorman's cruisers. This was doctrinally correct, and relied on the supposed superiority of Japanese gunnery to outrange the enemy.[74] The weather was perfect for shooting, and the Japanese had spotter aircraft overhead. Takagi was content to see what damage he could

MAP 5.1. BATTLE OF THE JAVA SEA, FIRST PHASE, 27 FEBRUARY 1942

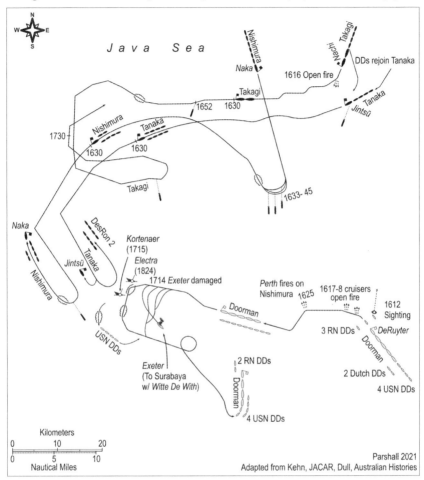

inflict during this phase, but his larger plan of battle was to "entice the enemy into a daytime engagement, and [then] destroy him at a stroke in a night engagement, making the most of the superior [Japanese] torpedo power"— precisely the tactical solution for which Japanese doctrine had called.

In the end, both cruiser forces ended up expending a great deal of ammunition to very little effect, although a solid hit by the Japanese cruiser *Haguro* on the British heavy cruiser *Exeter* temporarily disabled her. The Japanese also expended more than a hundred torpedoes, but sank only a single Dutch destroyer. Worse, as many as a third of their Type 93s detonated prematurely during their runs. "It was a true disaster," one Japanese officer recalled. "The operational units were so angry that they wanted the [Technical Department] personnel responsible to take their own lives."[75] It later developed that, in a case of pre-battle jitters, many of the ships had set their torpedo fuzes at too sensitive a setting. But this was hardly an auspicious beginning for one of Japan's most important weapons.

Having largely failed with daylight gunnery and torpedo attacks, Takagi settled down to a waiting game. Every time Doorman would try to break through to the convoy, Takagi would use his superior speed to force the Allied task force onto a heading away from his charges and to hold them at arm's length. In the process, Doorman lost another two destroyers, his ships were running low on fuel, and his crews were exhausted and frustrated. As darkness came down, the Japanese disengaged in order to reorganize their forces nearer the convoy and await developments.[76]

Doorman briefly considered retiring, but decided to make one last lunge toward the convoy. Takagi sighted the Allied force coming north in the gloom. Turning his cruisers parallel to Doorman's, but well ahead, he feigned retreat while actually setting Doorman up for a textbook torpedo shot.[77] At

MAP 5.2. BATTLE OF THE JAVA SEA, SECOND PHASE, 27 FEBRUARY 1942

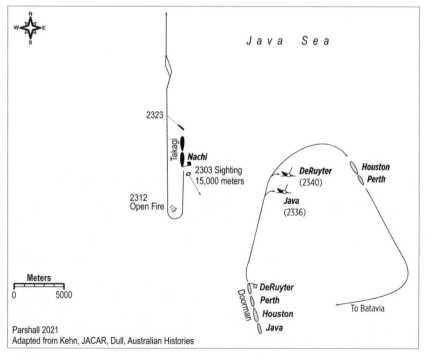

2323 Takagi's cruisers let fly over their right shoulders with a dozen Type 93s. To this point, the Japanese had fired 141 torpedoes, with only one hit. But this final salvo blasted first the Dutch cruiser *Java*, then Doorman's flagship *De Ruyter*. Both went down with extremely heavy loss of life, including Doorman. The battle had ended with finality. A total of five Allied warships and 2,300 sailors had been lost for just 36 Japanese killed.

In assessing the action, neither Japanese gunnery nor torpedo marksmanship was up to its high prewar expectations. Japanese post-battle reports particularly harped on the evils of wasting shells in long-distance gunnery duels—despite the fact that this had been a central tenet of prewar doctrine.[78] On the other hand, Takagi's basic stratagem of destroying his enemy "at a stroke" at night had been splendidly validated. Throughout the battles around Java, too, the Japanese were pleased to note that "The fighting ability of our special-type destroyers was superior and their reputation as the [Japanese navy's] principal weapon particularly in night engagements was confirmed."[79] On the other hand, the Japanese felt that Allied destroyers "seem to have had no intention to carry out aggressive torpedo attacks ([or] charges) either in the daytime and the night engagements, which gave us an impression that their power to conduct torpedo warfare is inferior."[80] They were not wrong. It would take more than a year for American destroyers to begin matching their opposite numbers.

The Battle of Savo Island

Five months after Japan's early-war triumphs came the wholly unexpected American invasion of Guadalcanal on 7 August 1942, which forced the Imperial Navy onto the defensive for the first time in the war. Their reaction was typically aggressive. Vice Admiral Mikawa Gunichi, pulled together a scratch force of five heavy and two light cruisers and a single destroyer and headed south from Rabaul to destroy the American invasion fleet off Guadalcanal.[81] This was hardly the Imperial Navy's A-team; only Mikawa's flagship *Chōkai* could be considered a modern warship, and the two light cruisers and destroyer were positively ancient. But Mikawa was determined.

On the Allied side, Rear Adm. Richmond Kelly Turner, the Guadalcanal invasion force commander, had adequate sighting reports of Mikawa's transit.[82] But he completely misjudged enemy intentions, and discounted the probability of a battle on the evening of 8 August. His tactical commander,

British Rear Admiral Victor Crutchley, issued instructions that did not allow for adequate concentration or coordination among the three separate task forces guarding the entrances to what would soon be known as Ironbottom Sound. Turner reviewed these dispositions and approved them. The result was a disaster.

Given the nature of his polyglot force, Mikawa opted for unorthodox simplicity, placing *Chōkai* in the lead, and his lone destroyer at the rear. The night was moonless and very dark, but Mikawa's superb lookouts sighted the American picket destroyer west of Savo Island without being detected by American radar—an event that was to happen to the Americans depressingly often in the months ahead. Hugging Savo's dark bulk, he slipped into the Sound and then sighted the Allied Southern Force some 12,500 yards away. It was 0136. Type 93s began hitting the water just two minutes later.[83] In accordance with Japanese navy doctrine, Mikawa held his gunfire until the last possible moment to give his torpedoes a better chance of hitting. The heavy cruisers *Chicago* (CA 29) and HMAS *Canberra* only detected Mikawa's force emerging from a cloud bank at 0143. Just seconds later, Mikawa opened fire, quickly

MAP 5.3. BATTLE OF SAVO ISLAND, 8–9 AUGUST 1942

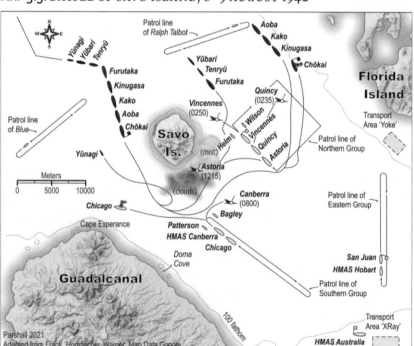

wrecking *Canberra* with two dozen shells and a possible torpedo hit.[84] *Chicago* was smacked by a Type 93 in her bow, and headed west out of the action, firing ineffectually. The Southern Force was thus destroyed in about seven minutes.

At 0146 destroyer USS *Patterson* (DD 392) had spread the alarm, signaling "Warning! Warning! Strange ships entering harbor." It did the Northern Force no good; Mikawa had already sighted them at the absurd distance of 16,500 meters even before he had opened fire on the Southern Group.[85] By the time Mikawa turned north to attack, though, his own formation had come apart into two groups, which were enveloping the Northern Force on either side. The three American cruisers in his sights—USS *Vincennes* (CA 44), USS *Quincy* (CA 71), and USS *Astoria* (CA 39)—had been slow to grasp the meaning of the rumblings that they had heard to their south. Their crews were exhausted, their skippers asleep, and it was only around 0145 that they began rousing themselves. Five minutes later, Japanese searchlights snapped on, pinning the American cruisers in their icy beams. Mikawa's gunners promptly poured in their fire. In just minutes, all three ships were heavily

PHOTO 5.4. A textbook example of using converging searchlight beams to illuminate a surface target. USS *Quincy* (CA 39), already heavily afire, is caught in the glare of one of Vice Admiral Mikawa's warships (probably *Aoba*). This is most likely just prior to the lights being flipped back off at 0155, at a range of around 6,200 meters. *U.S. Naval History and Heritage Command*

aflame, wrecked, and sinking. Mikawa swept past them, sprinkling more Long Lances, several of which duly found their marks. For all intents and purposes the battle was over by 0216.

Tactically speaking, Mikawa's onslaught had been somewhat atypical for the Japanese, in that he relied primarily on gunfire, guided by searchlights and star shells, as the instrument of decision. Since the crews in his scratch force had never trained together, he had ordered independent firing from the outset. It had all worked brilliantly. "The element of surprise worked to our advantage and enabled us to destroy every target taken under fire," Mikawa wrote later.[86] Turner and Crutchley had gotten their command blown out of the water—four cruisers gone and more than a thousand sailors lost. It was the worst defeat in the U.S. Navy's history.

With the battle won, though, a choice now lay before Mikawa—to turn around, seek out the invasion transports, and destroy them, or to retire. His force had suffered only minor damage but was badly dispersed and would take time to re-form. His chief of staff, too, was very concerned about what might happen if they were caught the next morning by American carrier aircraft.[87] Finally, Mikawa and his officers cannot have failed to be influenced by their indoctrination. They were the institutional product of a navy that had fervently worshiped for decades at the altar of the great naval thinker, American historian and strategist Alfred Thayer Mahan. In the Japanese navy's narrow interpretation of Mahan, it was believed that sea control devolved automatically from winning sea battles. Logistical matters were less important. Vice Mikawa had just won a great battle; sea control would therefore naturally devolve to the IJN whereupon the Imperial Army could finish off the American Marines at its leisure. Mikawa decided to retire. In retrospect, though, it probably would have been worth the sacrifice of every ship in Mikawa's squadron to destroy the American invasion transports that very night. Here was an example of the Imperial Navy winning a brilliant tactical victory, but not being able to place its actions correctly within a larger strategic context. The first—and probably best—chance for the Japanese to win the campaign at Guadalcanal had come and gone.

The Battle of Cape Esperance

As the fighting intensified on Guadalcanal, a naval squadron of three heavy cruisers and two destroyers under Rear Admiral Gotō Aritomo, was

MAP 5.4. BATTLE OF CAPE ESPERANCE

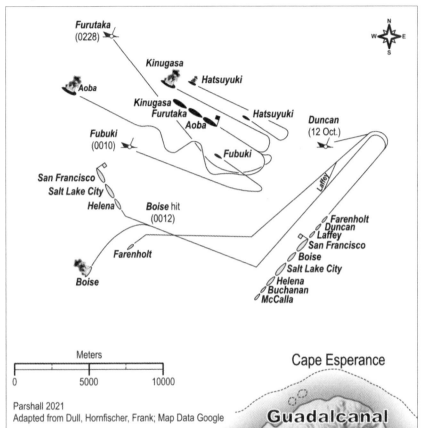

dispatched from Rabaul on 10 October to bombard Henderson Field and put it out of business.[88] Since the Japanese victory at Savo in August, the American theater commander, Vice Adm. Robert Ghormley, had been hesitant to commit surface forces to the area. But goaded on by Adm. Chester Nimitz, commander in chief of the Pacific Fleet, Ghormley was now rectifying that. Unknown to Gotō, that same evening the Americans had just landed reinforcements and had dispatched a task force of two heavy and two light cruisers, with five destroyers to cover the operation. This force, under the command of Rear Adm. Norman Scott, was now patrolling the entrance to Ironbottom Sound.

Scott had studied the disaster at Savo keenly, and felt that keeping his squadron concentrated and under tight control was the key to victory.

Accordingly, he deployed his ships in a single line, with his cruisers in the middle and destroyers on either end.

Luck was on the American admiral's side that evening, as the radar on light cruiser USS *Helena* (CL 50) began picking up the Japanese coming down the Slot at 2325 at a range of 28,000 yards.[89] Unwittingly, Scott had crossed Gotō's "T." For their part, Japanese lookouts did not discern the Americans until 2343, at a range of ten thousand meters. But Scott's flagship, USS *San Francisco* (CA 38), was deliberately not using her search radar, which Scott believed (wrongly) might be detected by the Japanese. His situational awareness suffered accordingly, and he hesitated to fire, compounding this with ambiguous communications to his task force. Finally, at 2346, with the range down to just five thousand yards, *Helena* would wait no longer. Flipping on her searchlights, she let loose with her formidable broadside of fifteen 6-inch guns. Her sister ship, USS *Boise* (CL 47) promptly followed suit.

Up to this moment, Gotō had refused to believe that the shadows ahead of him were enemy. He paid with his life. Even as the first American salvos rang out, heavy cruiser *Aoba* was still vainly blinking recognition signals at the Americans, assuming they were a Japanese resupply force that Gotō knew was a couple hours ahead of him. In response, the Americans landed immediate hits on *Aoba*'s superstructure, scything down the bridge watch and mortally wounding the admiral. With the darkness ahead now flashing with American gunfire, the Japanese formation quickly hauled about and began retreating. *Aoba* and *Kinugasa* survived, though both were heavily damaged. But their sister, *Furutaka*, interposing herself between Gotō's flagship and the American guns, was heavily hit and sank. Destroyer *Fubuki*, the original *Toku gata*, also went down. Even so, smart shooting by *Kinugasa* landed an underwater hit on *Boise*'s forward magazines that by all rights should have destroyed her. But luck and American damage control saved the day, and in the end Scott retired having only lost a destroyer.

After the debacle at Savo, Scott's victory was a welcome relief for the Americans. Linear tactics, coupled with radar, seemed to offer a solution to Japanese superiority in torpedoes and stealth. As it turned out, though, the victory was a tactical cul-de-sac. Linear formations actually offered a glorious target to Japanese torpedoes. And by shackling his destroyers to the movements of his cruisers, Scott had deprived some of his most capable night-attack vessels of the freedom of action they needed to close the range and attack effectively.

For the Japanese, this was their first taste of what it was like to be on the receiving end of radar-assisted gunfire. It had been most unpleasant. They had succeeded in spotting Scott at ten thousand meters—no mean feat in the dead of night. But they had been revealed to American radar at more than twice that distance, giving the Americans the priceless gift of time and initiative.

The First Naval Battle of Guadalcanal

By mid-November, things on Guadalcanal were at a crisis point for both sides. The Japanese army was starving in the jungle. American logistics also were far from good, and Vice Adm. William Halsey, the area commander, was down to just a single carrier. The Japanese, determined to run in a large supply convoy, decided to send a pair of fast battleships under Vice Admiral Abe Hiroaki, down from Truk to knock Henderson Field out so the convoy could make it through.[90] With Abe came a light cruiser and eleven destroyers. Waiting for them was a polyglot American task force of two heavy and three light cruisers, and eight destroyers under Rear Adm. Daniel J. Callaghan.

The Americans were keenly aware that they were massively outgunned. Although Callaghan issued no written instructions before the battle, the evidence strongly suggests that his intention (based on prewar doctrinal studies) was to get his heavy cruisers as close to the Japanese battlewagons as possible.[91] As far as the average American sailor knew, however, the coming encounter seemed a likely recipe for suicide.

Callaghan adopted a linear formation, selecting *San Francisco* as his flagship, despite her older radar. As at Cape Esperance, *Helena* picked up Abe coming into the Sound at a range of 27,000 yards. Abe's lookouts did not sight the Americans until he had closed to nine thousand meters. Both admirals remained in the literal dark as to the tactical picture. Precious seconds ticked away as the two formations kept closing each other at a combined speed of roughly thirty-eight knots (twelve hundred meters a minute). The result was that the Americans basically speared themselves into the middle of Abe's formation.

No accurate track chart of this battle has ever, or will ever, be drawn, because as soon as fire was opened at 0148 the tactical picture instantly devolved into complete chaos. It was, in words of one American skipper, "A barroom brawl after the lights had been shot out."[92] Both sides' ships maneuvered wildly, taking potshots at whatever crossed their sights. Enemy vessels reared up suddenly

MAP 5.5. FIRST BATTLE OF GUADALCANAL, 12–13 NOVEMBER 1942

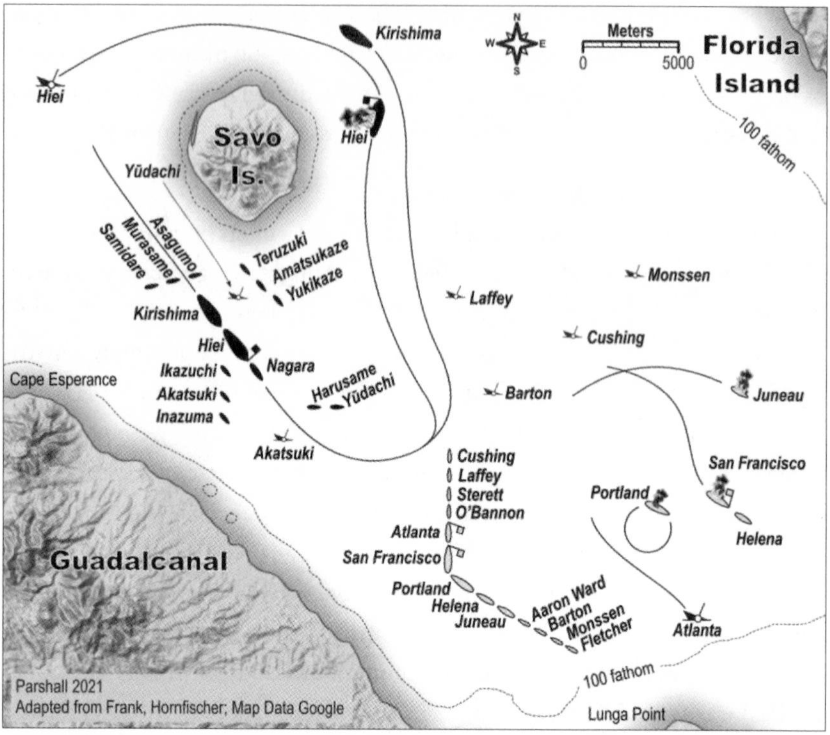

out of the darkness, exchanged frantic fire, then disappeared just as swiftly back into the gloom. Torpedoes crisscrossed the waters in a crazy quilt. Throughout the whole affair, the massive superstructures of battleships *Hiei* and *Kirishima* loomed, their searchlights probing the darkness, but attracting more than their share of fire from the Americans in return. *Hiei*'s upper works were soon streaming fire. In short order, Rear Admirals Callaghan and Scott were both killed—on the bridges of *San Francisco* and USS *Atlanta* (CL 51) respectively— Scott as the result of friendly fire. Despite heavy damage to *San Francisco* and USS *Portland* (CA 33), however, both American heavy cruisers were able to close, and *San Francisco* briefly exchanged fire with *Hiei* at point-blank range. Against all odds, she punched an 8-inch hole into *Hiei*'s steering compartment, knocking out her rudder. Abe, his force scattered, and his flagship burning and crippled, aborted the bombardment mission and began trying to disengage.

Five American warships were sunk—light cruiser *Atlanta* and four destroy- ers. *San Francisco*, *Portland*, and USS *Juneau* (CL 52) were heavily damaged.

Juneau would be lost later that day to a Japanese submarine. The Japanese lost two destroyers. *Hiei* could not be saved, either, sinking later on the 13th—the first Japanese battleship lost during the war. Callaghan's plan—communicated to no one—had worked. Though his squadron had been almost totally destroyed, it had accomplished its mission, at least for the time being.

From a doctrinal and tactical standpoint, a number of things stand out. First, the Japanese had been worsted by the Americans, who were using "Nikuhaku-hitchū!" tactics against Abe's Main Body—a bit of irony if ever there was one. Second, it is curious that the Japanese did not commit greater force to their effort. *Hiei* and *Kirishima*'s sisters, *Kongō* and *Haruna*, also were available, but they were not used. At this juncture, with the campaign hanging in the balance, and victory still available to either side, relying on half-measures represented a crucial mistake.

The Second Naval Battle of Guadalcanal

The fighting around Guadalcanal was approaching its climax. The night after Vice Admiral Abe's repulse, the Japanese bombarded Henderson Field with a pair of heavy cruisers, but did not inflict much damage. The Japanese navy was still determined to put Henderson out of business and run its supply convoy in.[93] Disgusted by Abe's withdrawal, his superior, Vice Admiral Kondō Nobutake, took matters into his own hands. Hoisting his flag in heavy cruiser *Atago*, he swept down from near Ontong Java Atoll (south of Truk) with *Kirishima*; *Atago*'s sister, *Takao*; two light cruisers; and nine destroyers. *Kongō* and *Haruna*, though, were again left behind.

Halsey was now in desperate straits. His battered squadron from the previous affray had withdrawn. There was nothing left in Ironbottom with which to fight Kondō, so he decided to strip the damaged carrier *Enterprise* of some of her escorts. Detaching fast battleships USS *Washington* (BB 56) and USS *South Dakota* (BB 57) and a paltry screen of four destroyers, he sent them north under Rear Adm. Willis A. Lee. Halsey was taking hair-raising risks, sending the last two American heavy units in the theater to just about the last place they should want to fight. The terrible danger of Japanese torpedoes—in constricted waters, at night—was self-evident. But Ironbottom Sound was small enough that a battleship could lob a shell clear across it. In essence, Lee was being sent into the nautical equivalent of a cage match, with few escape routes if things went wrong. Halsey, though, was determined to hang onto

Henderson Field, and he was willing to fight down to his last bloody rowboat to keep it.

Fortunately, in Lee he had found the perfect commander for a nighttime action. An Olympic gold medalist marksman, Lee had made his name as a gunnery specialist. Possessed of a phenomenal spatial awareness, he was also intimately familiar with the technical intricacies of *Washington*'s various radar sets and had drilled her gun crews to perfection. The stage was now set.

Lee edged into the waters off Guadalcanal carefully, looping around the north side of Savo, then heading into the heart of the Sound before turning back to the west toward Cape Esperance. As soon as he did so, his radars detected targets to the north at 18,500 yards. Though Lee did not realize it, Japanese lookouts from Kondō's eastern force, led by light cruiser *Sendai*, had had him under observation for half an hour already. Lee fired. The Japanese retreated. As soon as Lee began traversing the southern channel, though, disaster struck—a second Japanese light squadron had already fired torpedoes at him. Lee's four destroyers were quickly sunk or heavily damaged. Swerving to avoid the crippled destroyers, *South Dakota* found herself backlit by their pyres. Kondo's battle force quickly took advantage of this. Flipping on their searchlights, *Kirishima*, *Atago*, and *Takao* took *South Dakota* under withering fire. Though *South Dakota*'s watertight integrity was not damaged a whit, her superstructure was riddled by two dozen hits, and power failures kept switching off her radars. Blinded and afire, *South Dakota* wisely began making for the exits.

Throughout all this, Lee in *Washington* had remained shrouded by the black backdrop of Guadalcanal, quietly developing a fire-control solution against *Kirishima* using both optics and radar. When Japanese searchlights came on, the tactical situation was clarified beyond all doubt, and Lee ordered *Washington* to fire. The range was eighty-four hundred yards—not "point blank," as it often is described in accounts of the action, but definitely within what one of *Washington*'s officers described as "body-punching range."[94]

Washington's fire proved absolutely devastating, pulverizing *Kirishima* with as many as twenty 16-inch hits in just minutes.[95] Smashed and burning, *Kirishima* staggered out of line, already damaged beyond redemption. *Washington* continued heading northwest, feinting toward the Japanese reinforcement convoy, and thereby covering *South Dakota*'s retirement. Soon enough, though, Lee decided that discretion was the better part of valor in the face of a likely

MAP 5.6. SECOND BATTLE OF THE GUADALCANAL, 14–15 NOVEMBER 1942

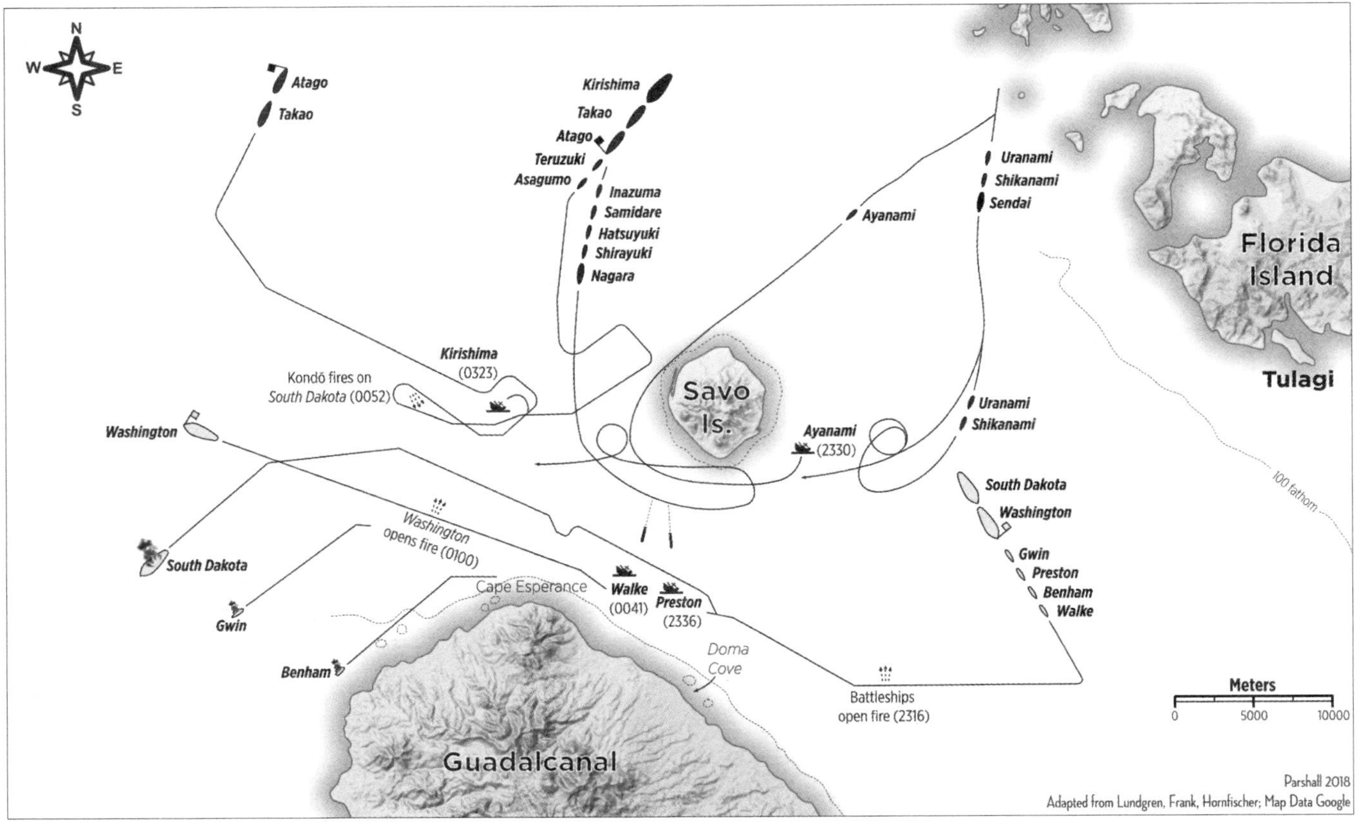

Florida Island

Tulagi

Savo Is.

Guadalcanal

Cape Espérance

Doma Cove

100 fathom

Uranami
Shikanami
Sendai

Ayanami

Uranami
Shikanami

South Dakota
Washington

Gwin
Preston
Benham
Walke

Ayanami
(2330)

Kirishima
Takao
Atago
Teruzuki
Asagumo
Inazuma
Samidare
Hatsuyuki
Shirayuki
Nagara

Atago
Takao

Kirishima
(0323)

Kondō fires on
South Dakota (0052)

Washington
opens fire (0100)

Washington

South Dakota

Gwin

Benham

Walke
(0041)

Preston
(2336)

Battleships
open fire (2316)

N
E
W
S

Meters

0   5000   10000

Parshall 2018

Adapted from Lundgren, Frank, Hornfischer; Map Data Google

torpedo attack, whereupon he began heading south himself. He was not wrong to be cautious, for the Japanese did indeed send schools of Type 93s swimming after his swiftly moving flagship. But their target angles were lousy, and although several fish passed uncomfortably close to *Washington*, Lee escaped. Japan's last chance to win sea control around Guadalcanal had evaporated.

It seems clear that strategic and operational errors had cost the Japanese their final chance at victory at Guadalcanal. At the tactical level, the Japanese navy had fought acceptably well, albeit it had been unlucky with its parting torpedo shots at *Washington*. But the action should never have come down to that. Had *Haruna* and *Kongō* come along to the party (let alone the battleships *Yamato* and *Mutsu*, which also were at Truk), even an exceptional leader like Lee would have been very hard put to it. But the Japanese were not willing to commit the units they needed to secure victory.

For the Americans, at the tactical level the superiority of their latest model search and fire-control radar was beginning to vindicate their prewar emphasis on gunnery. Then again, the U.S. Navy could hardly count on having battleships available for future engagements, or leaders as tactically skillful (and lucky) as Lee.

The Battle of Tassafaronga

Despite losing two battleships in two nights near Ironbottom Sound, the Imperial Navy was not quite ready to throw in the towel on Guadalcanal. Japanese troops were still starving there. For months now, the Japanese had been using fast destroyers (the famed "Tokyo Express") to run supplies down the Slot to the island. With the defeat of the resupply convoy in the wake of *Kirishima*'s demise, the Japanese had little choice but to continue their runs. On 29 November, the highly experienced Rear Admiral Tanaka Raizō brought his convoy of eight destroyers out of Shortlands.[96] Waiting for him was a powerful American task force under Rear Adm. Carleton H. Wright, comprising four heavy cruisers, one light cruiser, and six destroyers.

Wright possessed a crushing superiority in gun power. Unfortunately for the Americans, though, he had only been in command for two days, and was unfamiliar with the battle plan to which his squadron had been training. His destroyers ostensibly were slated to operate independently of the cruiser formation. As would be shown, however, Wright intended to keep them on a tight leash. Worse, the performance of the Long Lance still was not understood,

MAP 5.7. BATTLE OF TASSAFARONGA, 30 NOVEMBER 1942

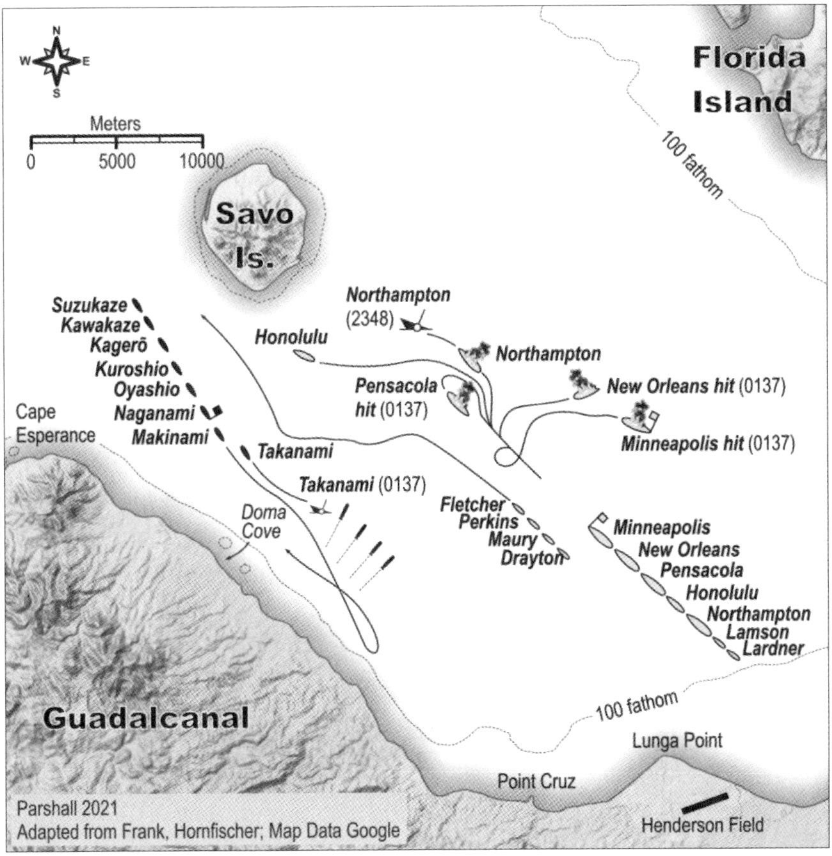

with the Americans assuming that ten thousand yards was probably a safe distance to operate from Japanese warships in order to avoid "torpedo water."

The evening of 30 November was moonless, overcast, and so incredibly dark that not even the Japanese lookouts could make out much. Starting at 2306, though, American radar began catching electronic inklings of Tanaka's force making its way into the Sound. Guadalcanal's black bulk still largely hid them from view. When the American destroyer division asked for permission to fire torpedoes at 2315, Wright declined, believing the range too great, thereby spoiling a beautiful firing solution. Tanaka had by now belatedly detected the Americans. He cancelled his supply drop and was preparing to attack.

Six more minutes passed before Wright's cruisers finally opened up, whereupon they quickly reduced destroyer *Takanami* to a wreck. Shifting fire to other targets, the Americans believed that the radar blips that they saw winking on and off were sinking vessels. But they were actually water spouts from their own shell splashes. Tanaka's sailors were unimpressed, remarking later that American gunnery "was inaccurate, shells [im]properly set for deflection were especially numerous, and it is conjectured that either [American] marksmanship is not remarkable or else the illumination from [their] star shells was not sufficiently effective."[97]

Throughout Wright's ineffectual fusillade, Tanaka's men held their own fire, calmly ranged on the American gun flashes, let fly with forty-four Long Lances, and then came about and ran for the exits. Wright's cruisers obligingly remained on a steady course, and were justly rewarded for their folly. Wright's flagship USS *Minneapolis* (CA 36) was blasted by two Long Lances, USS *New Orleans* (CA 32) lost her bow to another, USS *Pensacola* (CA 24) collected a hit and burst into flames, and USS *Northampton* (CA 26) received two hits astern. Only USS *Honolulu* (CL 48) escaped unscathed. *Northampton* later sank, and only superlative American damage control prevented the other three wounded cruisers from following her down. In the confusion, Tanaka made his retreat.

Tanaka had brilliantly demonstrated again the value of the torpedo in making stealthy counterattacks against an incautious enemy. Yet, he was not without his critics in the Imperial Navy, some of whom felt that he should have been flying his flag in the van (and thereby, presumably, shared *Takanami*'s fate).[98] These criticisms, though, seem well wide of the mark. Given the disparity in raw combat power between the two forces, and also the fact that Tanaka had been detected first, his victory looks all the more remarkable. It was a vindication of Japanese prewar training (emphasizing stealth, tactical flexibility) and excellent torpedo fire-control. On the American side, there was a recognition that its linear formations were dangerous, along with a growing realization that American destroyers should be given more freedom of maneuver.[99]

ASSESSMENT

The outcome at Tassafaronga offers an appropriate stopping point for this narrative, since it neatly juxtaposes a brilliant Japanese tactical victory set against a backdrop of strategic failure. Tassafaronga proved yet again what skillful, often brilliant night fighters the Japanese truly were. Their tactics were flexible, their crews and skippers aggressive, well-trained, quick to react, and accurate with their weapons. As an American officer candidly acknowledged, "The Japanese seem always to have their torpedoes in the water immediately."[100] The Americans had learned to both fear and respect those same torpedoes.

And yet despite this, the Imperial Navy now found itself at a crossroads. It was exhausted. Since August 1942, the fighting in the waters around Guadalcanal had cost the Japanese twenty-four major warships sunk, including two of their twelve battleships. Dozens more vessels had been damaged. The Americans, too, had suffered heavily, losing twenty-five warships of their own, including two fleet carriers and eight cruisers. But the Americans were still winning the battle for the island. Why was this?

There were two primary causes. First, although the Japanese had shown a propensity for technological innovation during the interwar period and were often flexible on the battlefield, they were oddly *inflexible* when it came to matters of doctrine and strategy. Not only that, but their commanders had difficulties placing their tactical actions within a larger strategic context. This had begun in the very first engagement at Savo Island, with Mikawa missing a crucial opportunity not just to defeat the enemy's warships, but to defeat an entire enemy campaign by attacking its logistical basis at the outset. As the fighting continued, it took longer than it should have for the Imperial Navy to shift its mental models regarding Guadalcanal's importance. This brutal, bitter struggle for a squalid malarial hellhole didn't look *anything* like what prewar Japanese navy doctrine would have called a Decisive Battle. But that's what it truly was. Yet from Admiral Yamamoto Isoroku, the commander in chief of Combined Fleet, on down, Japan's naval leaders found it difficult to accept that notion and then fight accordingly.

Even after having come to the decision in mid-October that it was worth using their old fast battleships to bombard Henderson Field, a month later the Japanese navy was still unwilling to commit *sufficient numbers* of those heavies to *win*. During the two crucial mid-November battles, *Haruna* and *Kongō* had both been available, but inexplicably the Japanese had not seen fit to send them along with either Abe or Kondō. Meanwhile, *Yamato* and *Mutsu* rode at anchor at Truk.[101] By contrast, Halsey fought for Guadalcanal with everything he had. And on the night of 14–15 November, he had been willing to bet his last nickel by sending in Lee's battleships with only the flimsiest of screens.

The second major problem for the Japanese was that key pieces of their prewar doctrine were failing. The central bet the Imperial Navy had made was that a high-quality few could conquer the mundane many. Superior technology, the Japanese believed, would provide the necessary edge to overturn the remorseless logic of Lanchester's formulas. Night combat would be the venue for effectively deploying that technological advantage. But those bets were not paying off. For instance, even though the Type 93 had proven itself a tremendous torpedo—the best in the world—it was still just a torpedo. Even good as it was, its accuracy was not what had been hoped. Nor was it enabling Japanese warships to outrange the enemy sufficiently to avoid having to absorb losses in return. Indeed, the constricted waters in the Solomons, with their shadowy island backdrops, often militated against long-range engagements. Thus, for the Japanese, even having the best torpedo in the world had not changed the broader parameters of the war all that much. At the same time, though, the excellence of Japanese night optics, flashless powder, and star shells were all being obviated by American improvements in a next-generation technology—radar—that *was* indeed changing the war.

The Japanese, of course, were working to play catch-up in deploying radar and radar detectors on their own warships. But critically, the Japanese navy forces also were late in making the conceptual leap to not just deploying these new sensors, but consolidating their data along with traditional radio, optical, and plotting inputs; developing techniques for synthesizing and interpreting that aggregate information flow; and then using it to manage night battle. In the end, it would be these higher-level techniques, which found their physical manifestation in things such as the U.S. Navy's Combat Information Center, that would prove critical to making the next leap in World War II night combat.

Some of the Imperial Navy's slowness in shifting its mental models may have stemmed from its inability to assess its own performance honestly. Where the Americans had been smacked in the face with several obvious, stinging defeats that demanded drastic remedial action, Japanese opinion regarding their own battlefield execution seems to have been far more sanguine. On 8 December 1942, just a week after Tassafaronga, Yamamoto's chief of staff, Vice Admiral Ugaki Matome, penned an entry in his diary commemorating the one-year anniversary of the start of the war, and the thousands of Japanese sailors who had given their lives so far in service of their Emperor.[102] In that time, naval general headquarters claimed, the Imperial Navy had sunk 11 American battleships, 11 carriers, 46 cruisers, 48 destroyers, and 93 submarines—a stunning total of 209 major warships.[103] Ugaki blandly commented that even if the list contained some errors, it still represented "a considerable amount." In actuality, to date the Americans had lost a total of 47 warships; the Japanese 51.[104] In the Solomons the ratio had been 25 Allied losses to 24 Japanese. For his part, Admiral Nimitz had not been slow in discerning this Japanese tendency to over-claim, writing his wife in the aftermath of the Battle of Santa Cruz: "I wish we had as many carriers as they claim to have sunk."[105] In other words, the Japanese were convinced they were trading losses at roughly 4:1, when the actual ratio was essentially parity. This was not a healthy trend if the Japanese were to evaluate the situation and adapt effectively. Unfortunately for the IJN in a contest between its *preferred* reality—and the *real* reality in the waters around Ironbottom Sound—the Ugaki diary version of things was not doing terribly well.

Indeed, just two weeks later, the Imperial Navy would finally come to the conclusion that the attrition being suffered in the bloody Guadalcanal campaign simply was not sustainable. The Imperial Army was in the same boat, having lost tens of thousands of men already on what was becoming known as "Starvation Island." Both services therefore agreed to begin an "advance to the rear" and abandoned Guadalcanal. The Japanese Empire's high-water mark had come and gone. But everybody knew that the savage night fighting in the Solomons was far from over.

CHAPTER 6

MASTERING THE MASTERS
THE U.S. NAVY, 1942–1944

TRENT HONE

O n the night of 17 March 1915, destroyers of the U.S. Atlantic Fleet Torpedo Flotilla approached a screened formation of "enemy" battleships. The destroyer captains were intent on proving the value of their training and attack doctrine. During previous exercises, they had raced through the battleship formation at full speed, using red flares to indicate the firing of torpedoes. The lack of live weaponry meant that many officers still believed "destroyers could not get within torpedo range without being discovered" and that "they could [not] make hits if they did."[1] On this night, one of the four destroyer divisions was allowed to fire actual torpedoes to demonstrate what the small ships could do.

After a conference on board light cruiser USS *Birmingham* (CS 2) to plan their attack, Lt. Cdr. Arthur Crenshaw led the destroyer captains against the "enemy." In the moonless night, they made an "entirely successful" attack against the screening vessels—hitting three with torpedoes—and made directly for the battleships. "Two torpedoes struck the leader, *Arkansas*, three struck the *New York*, one struck the *Texas*, and one the *Michigan*." Although the torpedoes were inert, the force of the impact was "very distinctly felt." Of eighteen torpedoes fired, eleven certainly hit. Two others might also have scored. If the three other divisions had also fired torpedoes, ninety of them—assuming a similar hit rate—would have found their mark.[2]

The destroyer commanders' claims were justified. Destroyers could penetrate screens without being seen. They could approach screening vessels undetected and ambush them. Once inside an enemy formation, the small ships could get "within easy torpedo range of battleships" and attack "with practicable certainty of success."[3] Searchlights were no defense; they only made the vessels using them better targets. The exercise proved that the U.S. Navy's destroyers had the potential to be lethal weapons of stealth and, when led resolutely, could make devastating attacks under cover of darkness.

However, by 1942, the Navy no longer was employing these effective tactics. As night-combat doctrine evolved in the interwar period (1919–1939), destroyer commanders emphasized alternative approaches. Guns became more important than torpedoes. Stealthy nocturnal attacks gave way to chaotic actions dominated by the exercise of individual initiative. These approaches reached their zenith in the battles off Guadalcanal during late 1942. Straining under the pressures of a global war, the Navy broke up the cruiser divisions and destroyer squadrons essential to effective indoctrination and relied instead on the skill and experience of individual commanding officers and their crews. It proved enough to succeed at Guadalcanal—barely—but it was insufficient for long-term success.

In 1943 the Navy transformed its night-combat tactics. Increasing stability in squadrons and divisions permitted the creation of coherent plans and doctrine. The introduction of Combat Information Centers (CICs) and effective plotting techniques gave ship and formation commanders improved situational awareness, enabling them to capitalize on the potential of radar and coordinate the actions of distributed formations. The importance of stealth—as demonstrated by the Atlantic Fleet Torpedo Flotilla—was recognized once more. The new tactics eventually overcame the advantages of the Imperial Japanese Navy and enabled the U.S. Navy to begin to dominate night surface actions in the Pacific. However, it took time for the Americans to embrace the potential of the new techniques.

NIGHT SEARCH AND ATTACK

The aggressive doctrine devised by the Atlantic Fleet Torpedo Flotilla became known as the "Night Search and Attack." First issued in a tentative form in November 1913, the doctrine introduced a new paradigm in the Navy. From then on, doctrine and plans would be foundational elements of the Navy's approach to tactics. Doctrine provided high-level concepts for how particular forces were to operate. In the "Night Search and Attack," for example, destroyers would seek out, locate, and then attack a screened enemy formation; that was the doctrine. Before battle or an exercise, commanding officers would come together in conference to formulate the plan, agreeing how to tailor their doctrine and employ it for maximum advantage. Doctrine and plans worked together to allow Navy officers to translate their theories of war into coherent action.

These concepts emerged in the Atlantic Fleet Torpedo Flotilla because of the challenge of coordinating distributed formations of destroyers in night combat. They could not exchange detailed instructions in battle. Radio was unreliable and lights would give their position away. The commanding officers had to be "welded into a body, whose various members . . . can act collectively as a unit free from embarrassing internal friction."[4] When Capt. William S. Sims assumed command of the Flotilla in July 1913, he deliberately explored how to use doctrine to coordinate the actions of his ships. Lt. Cdr. Dudley Knox, the Navy's foremost advocate of the potential of doctrine, served as Sims' aide.

Knox "played the largest part" in developing the Flotilla's doctrine.[5] As he explained in his prizewinning 1915 *Proceedings* essay, it was impossible for a commander to have "cognizance of events immediately" because "his communication system [would be] too precarious and slow." Therefore, it was "imperative" that subordinates "decide and . . . act" on their own initiative. With doctrine, they could become "almost literally of one mind with their commander-in-chief and with each other." Thorough preparation and practice would create an environment of "frictionless and automatic teamwork."[6] In the Flotilla, Knox and Sims demonstrated the effectiveness of this approach and familiarized many young officers with it, including future admirals Ernest J. King, Harold Stark, Aubrey Fitch, Frank Jack Fletcher, and William F. Halsey Jr. Doctrine and plans became the basis of Navy tactical doctrine.[7]

UNCHALLENGED ASSUMPTIONS

During the interwar period (1919–1939), the Navy's approach was refined by preparations for a war in the Pacific against the Empire of Japan. The Night Search and Attack became a component of a broader objective—coming to grips with the Japanese fleet and defeating it in battle. Night Search and Attack exercises drove increasingly sophisticated mechanisms for defending the Navy's battle line at night, including enhanced screening practices, very-high-frequency radios for the rapid exchange of information, and improved fire-control procedures and equipment. At the same time, these exercises fostered more complex means of overcoming those defenses and mounting a successful attack. The Navy's aggressive destroyer tactics led to new techniques for both offense and defense. However, in the 1930s this interplay ultimately led the Navy away from the stealthy attacks pioneered in the Torpedo Flotilla.

The shift was visible in the Fleet Problems, contested exercises that explored operational scenarios likely to occur in a war against Japan. At the start of the 1930s, night destroyer attacks resembled those of 1915. In Fleet Problem XI of 1930, for example, USS *Concord* (CL 10) and five destroyers of the "Blue" fleet stealthily penetrated the "Black" screen and made a devastating attack on the enemy battle line.[8] Two years later, in Fleet Problem XIII, destroyers sank the aircraft carrier USS *Saratoga* (CV 3) and damaged her escorts. The commanding officer of USS *Williamson* (DD 244), Lt. Cdr. Aaron S. Merrill was "impressed" with the effectiveness of the attack. He felt that having individual squadrons strike independently had contributed to success because "the destroyer once discovered is doomed."[9]

Navy destroyer designs from this period reflected the "widespread belief that torpedoes were the primary destroyer weapons." In January 1933, the

PHOTO 6.1. USS *Craven* (DD 382) off the Mare Island Navy Yard in November 1943. *Craven* was a *Gridley*-class destroyer armed with sixteen 21-inch torpedo tubes, reflecting the Navy's emphasis on large torpedo armaments in the early 1930s. She was part of Cdr. Frederick Moosbrugger's Task Group 31.2 at the Battle of Vella Gulf. Note the searchlight amidships and lack of gunhouses for her aft 5-inch guns. *National Archives, 19-N-57249*

General Board specified that the eighteen ships of the *Mahan* and *Dunlap* classes would be armed with twelve 21-inch torpedo tubes. Older Navy destroyers also carried this many, but no other navy's destroyers did. By 1935, the Navy was planning even larger torpedo armaments. The next twenty-two ships—of the *Gridley*, *Bagley*, and *Benham* classes—had two quadruple torpedo mounts amidships on each side. With "curved ahead fire"—a creative use of gyro settings that made torpedoes quickly change course to their desired trajectory after they were launched—these destroyers could fire all sixteen torpedoes at a target in "a single massive salvo."[10]

However, the Fleet Problems also demonstrated that destroyers had difficulty finding enemy formations. In some exercises they cooperated with patrol planes in a "joint search and attack," but it was far more common to pair destroyers with the new heavy cruisers, built to the constraints of the Washington Treaty of 1922. The heavy cruisers were excellent scouts, and, as screening tactics became more effective, cruisers joined destroyer attack units. The cruisers provided "gunnery support," brought "superior force to bear on the enemy screening vessels," and helped the destroyers break through the screen.[11] In Fleet Problem XVIII of 1937, the "White" fleet formulated detailed plans for such attacks, and, although circumstances prevented their use, they were "recommended for further investigation."[12] These developments mirrored the Japanese tactics described in Chapter 5, which increasingly emphasized powerful night-attack formations that included heavy cruisers and battleships.[13]

One of the reasons for this shift was the increasingly powerful armament of the Navy's light cruisers. Built to the restrictions of the London Treaty of 1930, the ships of the *Brooklyn* class began to join the fleet in 1937. They were armed with fifteen 6-inch guns, each capable of firing as many as thirteen times per minute.[14] Designed to defend the battle line from torpedo attack, a single *Brooklyn* could overwhelm an entire destroyer division. In Fleet Problem XIX of 1938, for example, all the destroyers of the "Western" fleet's right flank were sunk attempting to make daylight torpedo attacks on the enemy battle line. Cruiser support was becoming essential to successful destroyer attacks, especially during the day, but also at night.

By 1940, the Navy had met this need with two new Night Search and Attack formations: the "V" and the "Wedge." Both integrated cruisers and destroyers together. The Wedge placed a line of cruisers ahead of two or more destroyer columns. It was developed for penetrating thin screens and designed to allow the attacking ships to pass through the enemy formation. The V, in contrast, anticipated a turn away. It placed two cruiser columns at the head of the V and destroyers at the base. On contact, the V would split, with each arm of the V turning to one side of "the enemy formation, raking it thoroughly at short range with cross fire [*sic*] of guns and torpedoes." Both the V and the Wedge anticipated using cruiser gunfire to pave the way for destroyer torpedo attacks; the stealthy tactics of a decade before had largely been abandoned.[15]

Even if they lacked cruiser support, Navy destroyer commanders were determined to fight their way past enemy screens. Tactical instructions encouraged ships to open fire on enemy screening vessels immediately and "turn searchlights on [enemy] . . . bridge and fire control stations."[16] By using their guns and searchlights, attacking destroyers would give their position away and be forced to make their torpedo attacks while under fire. Torpedo-firing exercises accounted for this by having targets flash blinker lights to simulate enemy gunfire. Stealth had become a secondary consideration; what mattered was a sufficiently violent attack with guns and torpedoes.[17]

PHOTO 6.2. USS *Charles Ausburne* (DD 570) was Capt. Arleigh Burke's flagship during the battles of Empress Augusta Bay and Cape St. George. She was a *Fletcher*-class destroyer, built to serve multiple roles, including fleet air defense, antisubmarine warfare, and night search and attack. She is pictured off Boston in March 1943. Note that her forward director is tracking the photographing ship. *National Archives, 19-N-41529*

New destroyer designs accounted for this. As the 1930s progressed, the Navy recognized that a war in the Pacific would require a sustained campaign. The Japanese navy would make repeated attritional attacks with light forces, submarines, and aircraft. Destroyer designs became more balanced, leading to multipurpose vessels that could screen the fleet and protect against these various types of attack. The new destroyers had dual purpose main batteries for use against airplanes, increased antisubmarine capabilities, and more light antiaircraft armament. The 175 ships of the *Fletcher* class designed in early 1940 were built to this model. They served extensively in the Pacific from late 1942 and were initially armed with five 5-inch guns, ten 21-inch torpedo tubes, depth charges, and adequate topside space for many antiaircraft guns.[18]

IRONBOTTOM SOUND

Throughout the interwar period, as the Navy explored how best to integrate new technologies, platforms, and capabilities, it fostered deliberate learning. The open-ended structure of the Fleet Problems encouraged experimentation with new doctrines, tactics, and techniques. Officers reviewed and assessed the results of these experiments from a wide range of perspectives to identify the most effective approaches and integrate them into the fleet. This process allowed the Navy to explore new ways of fighting and exploit the best of them for future use. In World War II, the Navy continued to use this process to identify lessons and develop new tactics rapidly.[19]

The battles fought off Guadalcanal in late 1942 were a crucial part of this effort. Following their prewar training, task force commanders experimented with new tactics to try to gain an advantage. They abandoned complex tactical forms such as the V and the Wedge and instead concentrated their ships in linear formations that they hoped would ease maneuvering and prevent friendly fire. Introduced by Rear Adm. Norman Scott, who considered it "most practical for night action," the linear formation appeared to work well at the Battle of Cape Esperance.[20] However, it was unwieldy for large formations, failed to prevent friendly fire, and was abandoned after the battles of mid-November in which Scott and Rear Adm. Daniel J. Callaghan were killed.

When Rear Adm. Thomas C. Kinkaid assumed command of the South Pacific's cruiser-destroyer striking force later that month, he developed a new plan. Kinkaid's first command had been destroyer USS *Isherwood* (DD 284) in 1924 and he knew what destroyers could do. He planned to send his destroyers ahead to make a radar-directed torpedo attack and "obtain the maximum benefits of surprise." Once the destroyers had completed their attack, the cruisers would open fire "at a range . . . between 10,000 and 12,000 yards."[21] Like the Navy's plans for daylight action, Kinkaid's approach coordinated the action of multiple types to maximize their effectiveness. His plan was a significant departure from prewar tactics, but an appropriate response to the battles off Guadalcanal. Unfortunately, before Kinkaid could employ his plan, he was detached from the South Pacific. Rear Adm. Carleton H. Wright tried to use it at the Battle of Tassafaronga, but he lacked familiarity with his subordinates, unduly constrained their initiative, and suffered a crippling defeat.[22]

One of the challenges Wright faced was that his cruisers used "continuous fire . . . instead of full gun salvoes."[23] Scott had endorsed this technique, which was a departure from the prewar practice of firing ranging ladders at night to establish the "hitting gun range" and then rocking ladders—adjusting the range between salvos to "walk" the shells back and forth across the target—to secure the maximum number of hits. Salvo fire made observing these ladders and bringing them onto the target much easier. However, because cruisers were now equipped with the FC (Mark 3) fire-control radar, Scott, Wright, and other officers believed they could fire their guns at the correct range continuously, without the need for ladders or salvos. They were incorrect, but Navy officers would not realize it until well into 1943.

Other experiments explored how best to maximize the potential of radar. A most promising experiment was conducted on board USS *Fletcher* (DD 445) in November 1942. Her captain, Cdr. William R. Cole, and his executive officer, Lt. Cdr. Joseph C. Wylie, recognized that the new microwave surface search radar, the SG, and its plan position indicator (PPI) display could provide much greater situational awareness than the view from the bridge. The two men devised a prototypical CIC. Wylie stood at the edge of the radar room, observed the radar screen, and kept Cole "continuously advised of the tactical situation."[24] Wylie also tapped into several of *Fletcher*'s communication circuits and coached the ship's weapons onto various targets. The two men's creative collaboration enabled *Fletcher* to survive the melee during the First Battle of Guadalcanal unscathed.

Fletcher's experiment addressed a need that also had been identified at Pacific Fleet headquarters. After assessing action reports, Adm. Chester W. Nimitz, the fleet commander, and his staff realized that ships were not making the best use of radar, and on 26 November he ordered every ship in the fleet to establish a CIC. Nimitz argued that "maximum combat efficiency . . . can best be attained through full utilization of all available sources of combat intelligence" and that radar had to be used to "maximum effectiveness." Accordingly, each ship was to establish a center with plots that would "maintain [a] continuous information summary" to keep "flag, ship, and fire control stations" updated.[25] Nimitz's instructions called for tying together a set of existing techniques and integrating them into a cohesive whole. He deliberately did not specify how best to do so; instead, he built on the Navy's prewar learning approaches. As each ship created and tested its CIC, more was learned about how to organize and operate them, allowing the best CIC techniques to be rapidly identified and promulgated throughout the fleet.[26]

At the same time, Nimitz issued new instructions for Night Search and Attack that reflected the war's lessons. In contrast to the prewar emphasis on "gun action," he stressed the advantage of stealth and surprise. Destroyers should "remain completely dark as long as possible" in order to "close, track [the enemy] by radar, and launch torpedoes before they are observed." An "ideal" squadron attack would involve three destroyer divisions approaching "from widely separated sectors" and hitting the enemy "with torpedoes before being detected." Like Kinkaid, Nimitz emphasized using radar—and the potential offered by the CIC—to coordinate the actions of distributed formations.[27]

One of the most important lessons of the Guadalcanal battles was that ships were shifting between task forces too frequently to develop the doctrine and plans so essential to coordinated action in combat. *Secret Information Bulletin No. 5*, part of a series of assessments issued by Adm. Ernest J. King, the Navy's commander in chief, observed that it was "unsound and a waste of material to throw forces together just prior to an action with no opportunity for [the] OTC [officer in tactical command] to issue instructions, doctrine, orders, etc." The inability of commanders to indoctrinate their forces meant that formations disintegrated on contact with the enemy, leaving ships to fight individually. As a result, the bulletin concluded, "We are paying heavily for this."[28]

NEW PLANS AND TACTICS

In 1943, as the Navy adapted to the hard-earned lessons of Guadalcanal, it initiated an offensive toward the Japanese stronghold at Rabaul. Adm. William F. Halsey Jr., the commander of the South Pacific Area, was determined to dominate the seas and skies and inhibit the movement of Japanese troops and reinforcements in the Solomons. Two cruiser-destroyer groups were his main surface striking forces. One was commanded by Rear Adm. Warden L. Ainsworth, and the other by Aaron S. Merrill, now a rear admiral. Both developed tactics based on Nimitz's new guidance from late 1942.[29]

Merrill was first to put the new tactics into practice. Halsey sent Merrill's Task Force (TF) 68 to bombard Japanese installations at Vila from inside Kula Gulf on the night of 5 March 1943 and ordered him to "destroy" any enemy ships he encountered.[30] Merrill held a conference with his commanding officers to review intelligence, prepare plans, and arrange for aerial support. Radar had created new possibilities for coordinating patrol planes and surface ships at night, and radar-equipped "Black Cats," a variant of the PBY Catalina, regularly patrolled the Solomons. One would spot for Merrill's bombardment.

During the approach, the Black Cat sighted what appeared to be two enemy cruisers. Merrill prepared for a fight. He sent USS *Waller* (DD 466) ahead as a radar picket. Under the direction of then–Cdr. Arleigh A. Burke, commander of Destroyer Division 43, *Waller* swept the waters of the gulf from a position six thousand yards ahead of Merrill's cruisers. Four thousand yards behind her, USS *Conway* (DD 507) screened the cruiser line—USS *Montpelier* (CL 57), USS *Cleveland* (CL 55), and USS *Denver* (CL 58)—while USS *Cony* (DD 508) guarded the port quarter of the cruisers against enemy patrol boats that might appear out of the "many small coves which indent the eastern shore of the Gulf."[31]

The night was very dark. Merrill's ships navigated into Kula Gulf using their SG radars. Because each ship had only one, that presented a problem. They could not simultaneously develop accurate fixes on potential targets and navigate. Fortunately, one of the navigation points was Sasamboki Island. At 0057, it appeared on the correct bearing, but nine thousand yards closer than anticipated. A minute later, radar operators realized that the "island" was an unidentified ship. It quickly resolved into two contacts, approaching on a nearly parallel, but opposite course. Merrill told his ships to stand by to open fire. Burke had *Waller* launch a half-salvo of five torpedoes using radar control.

Shortly after 0101, with the range down to ten thousand yards, Merrill's cruisers opened fire.[32]

The second enemy ship in line presented the best radar target. It was Japanese destroyer *Murasame*. A series of hits set her ablaze and a "violent explosion" from one of *Waller*'s torpedoes sank her. The leading enemy ship, destroyer *Minegumo*, was now under the concentrated fire of Merrill's three cruisers. Numerous shell hits soon brought her to a stop. Merrill had *Waller* finish her off. The brief exchange once more proved the crucial importance of surprise. Merrill's ships were undamaged and *Murasame*'s lookouts had not seen them until they opened fire. Merrill correctly concluded that the Japanese were hit so quickly that they were unable to respond effectively.[33]

One reason Merrill's cruisers had fired so accurately was because they were equipped with the new FH centimetric fire-control radar, more commonly known as the Mark 8. The Mark 8 had a display similar to the PPI of the SG, but instead of being centered on the radar, the Mark 8's display was centered on the target. Under normal conditions, the target ship and the relative positions of nearby shell splashes could be distinguished, allowing the fire-control team to observe the fall of shot and correct their aim. To make this process more effective, Merrill's flagship, *Montpelier*, fired salvos. Merrill noted that the Mark 8 was "greatly superior to any other type of fire control radar" for distinguishing targets and spotting the fall of shot.[34]

Merrill's victory seemed to prove the value of the new tactics that he and Ainsworth had formulated. Merrill used destroyer torpedoes and cruiser gunfire together to successfully surprise and overwhelm a pair of enemy destroyers. However, the circumstances were unusual, and there were flaws. In his comments on the action, Halsey presciently noted that "both surface targets were not taken under fire simultaneously" and that in the future, "this might produce serious consequences."[35]

The Japanese would soon demonstrate the wisdom of Halsey's comment. They had also adjusted their tactics and were intent on maximizing the potential of their most effective night-combat weapon, the Type 93 torpedo. U.S. Navy officers remained ignorant of the Type 93's capabilities well into 1943. There were, however, some important clues as to its potential. On the night of 4 July 1943, Ainsworth's task force entered Kula Gulf to bombard Japanese positions at Vila and Bairoko Harbor. The bombardment covered the landing of troops at Rice Anchorage, part of Halsey's campaign to seize New Georgia.

PHOTO 6.3. The ships of Rear Adm. Aaron S. Merrill's Task Group 36.2 bombard Japanese positions at Munda on the island of New Georgia in the early morning hours of 12 July 1943. The smoke, bright flashes, and tracers from their gunfire contrast against the dark night. U.S. Navy ships repeatedly bombarded Japanese positions at Vila and Munda during the New Georgia campaign. They also fought several night battles. The night after this bombardment, Rear Adm. Warden L. Ainsworth's Task Group 36.1 fought the Battle of Kolombangara. *National Archives, SC 181667*

Destroyers USS *Nicholas* (DD 449) and USS *Strong* (DD 467) were leading Ainsworth's formation, sweeping for potential threats. As the cruisers finished their bombardment of Bairoko, *Strong*'s gunnery officer, Lt. James A. Curran, saw a torpedo wake approaching from the port side. The impact tore a large hole in the hull and eventually broke the destroyer in two. Ainsworth had anticipated that submarines might interfere with his bombardment, and since no radar contacts were made within the gulf, he concluded that *Strong* had been the victim of an enemy submarine.[36]

However, Capt. Francis X. McInerney, who commanded Ainsworth's destroyers, thought "enemy destroyers might have been present" and fired the torpedo that sank *Strong*. Before the torpedo hit, USS *Ralph Talbot* (DD 390) had picked up two radar contacts exiting the gulf.[37] These were part of a group of four Japanese destroyers, led by *Niizuki*, which had a prototype radar. The radar alerted Rear Admiral Akiyama Teruo that there were U.S. ships inside Kula Gulf. He abandoned his planned reinforcement mission but fired torpedoes into the gulf before turning away. One of them found *Strong* after a run of some eleven miles.[38] *Secret Information Bulletin No. 10* concluded it was possible that a Japanese destroyer had fired the weapon from a concealed position in the radar shadow of Kolombangara Island.[39]

Before the implications of *Strong's* loss were fully recognized, Ainsworth would fight the Battle of Kula Gulf. On 5 July, Akiyama was on his way back to Vila with ten destroyers and more reinforcements. Halsey sent Ainsworth to intercept. Ainsworth's plans for night battle referenced the "general principles" of the new Night Search and Attack doctrine and emphasized exploiting surprise "to the fullest extent." However, Ainsworth planned to keep his force concentrated to "bring its full strength against the enemy." That meant using the "superior volume of fire" of his cruisers. On dark nights, Ainsworth intended to open fire "beyond the maximum range of visibility . . . at medium ranges of 8,000–10,000 yards" and use flashless powder to keep the position of his ships obscured. If there was sufficient visibility, Ainsworth planned to open fire from 13,000 yards, using star shells to illuminate the enemy and augmenting his fire-control radars with visual observations. In both plans, torpedoes were largely an afterthought.[40]

The night was "very dark, no moon, overcast, [and with] passing showers." At 0136, Ainsworth's flagship, USS *Honolulu* (CL 48), picked up contacts with her SG radar 20,500 yards away, off the coast of Kolombangara. Five minutes later he ordered his ships into battle formation and instructed them to use the plan for dark nights. As Ainsworth tracked Akiyama's ships and maneuvered into firing position, he noted that the Japanese were in two groups. He issued a series of contradictory orders that confused McInerney. At 0153, McInerney thought he received an order to commence firing; that was not part of the plan, so he asked for clarification. Ainsworth told him to wait and clarified that the cruisers would fire on each enemy group in turn. This exchange and Ainsworth's apparent uncertainty suggests that he was personally observing the SG display and not using the enhanced information processing capabilities of his CIC. By the time Ainsworth had determined how to fight the battle, the range was down to seven thousand yards.[41]

Several minutes earlier, at 0147, lookouts aboard Akiyama's flagship, *Niizuki*, sighted Ainsworth's cruisers. Akiyama ordered his three leading ships—which were serving as escort to the transport destroyers—to concentrate and make a torpedo attack, but at 0157, before *Niizuki* could launch torpedoes, Ainsworth's cruisers started shooting. All three of them quickly scored hits on Akiyama's flagship. *Honolulu* and USS *St. Louis* (CL 49) were relatively obscured because of their flashless powder. USS *Helena* (CL 50) had expended almost all of hers in the bombardment the night before; her gun flashes made her a good target. *Suzukaze* and *Tanikaze* unleashed a salvo of Type 93s before reversing course, making smoke, and opening fire with their guns. *Niizuki* was doomed; *Suzukaze* was damaged by four hits, but relatively unharmed; *Tanikaze* was struck by a single dud. Ainsworth incorrectly assessed that "this first group of enemy vessels ... [was] practically obliterated by the tremendous volume of fire."[42]

At 0203, Ainsworth ordered a reversal of course so that he could attack the other enemy group. Before *Helena* could execute the turn, three torpedoes struck her in relatively quick succession, shearing off her bow and breaking her hull amidships. Another torpedo hit *St. Louis*, but failed to detonate. One more barely missed *Honolulu*. With both formations confused and disrupted, Ainsworth "fired as targets of opportunity presented themselves." Akiyama's second group, the seven transport destroyers, engaged, but Ainsworth's gunfire dissuaded them from closing. Four shells hit *Amagiri*; others damaged *Hatsuyuki*'s steering; and one struck *Nagatsuki*. The transport destroyers withdrew into the shadow of Kolombangara. Ainsworth once more overestimated the damage inflicted. He claimed sinking a total of six ships, but only *Niizuki* and *Nagatsuki* were lost, the latter because she ran aground.[43]

Part of the reason for Ainsworth's inaccurate assessment was that his cruisers used the older FC fire-control radar. Unlike the Mark 8, the FC used an A-Scope display based on an oscilloscope. Range was presented along the horizontal axis and strength of the radar return signal on the vertical; stronger returns produced taller "pips." When using the A-Scope, it was easy to lose the target among the radar returns generated by shell splashes, especially when firing continuously, as Ainsworth's ships did. The challenge was exacerbated by their fire-control procedures. To spot the fall of shot, FC operators focused their equipment onto a narrow range band, and, if an enemy ship changed course, it could quickly disappear from the screen. When this happened, the target was erroneously claimed as "sunk."[44]

Helena's loss was another valuable clue to the capabilities of Japanese torpedoes. Ainsworth's plot of the action showed that if Akiyama's ships had "placed torpedoes in the water at the exact moment we opened fire, these

MAP 6.1. BATTLE OF KOLOMBANGARA, 12–13 JULY 1943

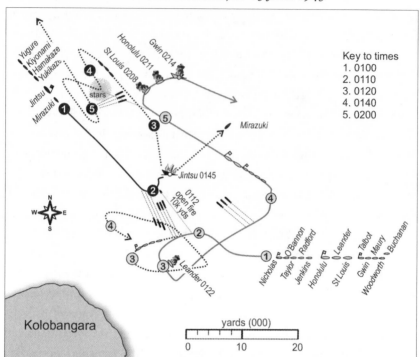

torpedoes would have to be faster than any we possess in order to reach the *Helena* at the time she was hit." Evidence was mounting that the Japanese navy had torpedoes of unprecedented speed and range. Although Ainsworth was not ready to draw that conclusion, he did offer that the Navy had "much to learn from them [the Japanese] about effective torpedo fire."[45]

That would become even more evident a week later at the Battle of Kolombangara. On 12 July, Halsey sent Ainsworth to intercept another "Tokyo Express" run. To augment Ainsworth's task force, Halsey ordered his amphibious force commander, Rear Adm. Richard K. Turner, to give Ainsworth "all available destroyers." That gave Ainsworth three cruisers—his flagship, *Honolulu, St. Louis,* and HMNZS *Leander*—and ten destroyers, five in the van under Captain McInerney and five in the rear under Capt. Thomas J. Ryan. The six destroyers that joined Ainsworth were from three separate squadrons; they were unfamiliar with each other and with Ainsworth's tactics. The Navy was still finding it difficult to employ cohesive formations even though the importance of effective indoctrination was understood. Indeed,

Ainsworth himself noted that "destroyers must be trained as a unit . . . to be of value in battle"—the Navy was still finding it difficult to employ cohesive formations.[46]

Scouting seaplanes kept Rear Admiral Izaki Shunji apprised of Ainsworth's approach. His flagship, light cruiser *Jintsū*, and five escorting destroyers were covering four destroyer transports. Anticipating a fight, Izaki detached his destroyer transports and sent them ahead to Sandfly Harbor. At 0057, he received a report from a plane that enemy ships were approaching. However, Izaki's lookouts did not sight Ainsworth's ships until 0108. By that time, Ainsworth had already put his plan in motion. His radars had located Izaki's formation at 0059, range 30,000 yards. Ainsworth sent McInerney and his van destroyers ahead; at 0110, they made a torpedo attack from about ten thousand yards. But Izaki had fired first; he turned away, avoided the American torpedoes, and illuminated Ainsworth's van destroyers with *Jintsū*'s searchlight. They opened fire. Ainsworth's cruisers, now on a southwesterly course, followed suit at 0112, from just over ten thousand yards.[47]

"At the end of five minutes," Ainsworth reported, "the three leading enemy ships were smoking, burning, and practically dead in the water." Once again, he had overestimated the effectiveness of his cruisers' gunfire. They concentrated on the best radar target, Izaki's flagship, *Jintsū*, and "smothered" her with more than twenty-five hundred shells. No other Japanese ships were hit, but *Jintsū* was doomed. After Ainsworth's cruisers opened fire, his rear destroyers unleashed a torpedo salvo. One of their torpedoes might have hit *Jintsū*, but it is unlikely. McInerney's destroyers fired additional torpedoes later in the action and one of them struck at 0145, guaranteeing *Jintsū*'s demise.[48]

The moon, already low on the horizon, became obscured by clouds. Visibility had been "about five miles." Now it was much less. At 0117, Ainsworth reversed course, intent on making another run past the enemy formation with his cruisers, but not all his ships received the order. Smoke from their gunfire, combined with the sudden darkness, disrupted the formation. As ships tried to regain or maintain their stations, Japanese torpedoes swarmed among them. One hit cruiser *Leander*'s port side, flooding a boiler room and knocking out all communications.[49]

While the old destroyer *Mikazuki* stood by *Jintsū*, Captain Shimai Yoshima took his four destroyers—*Yukikaze*, *Hamakaze*, *Kiyonami*, and *Yūgure*—north into a rain squall and commenced reloading torpedoes. A Black Cat observed them and incorrectly reported that they were retiring at high speed. Ainsworth, who had been "looking at the PPI over the shoulder of the radar operator and directing the operator where to train," began to lose situational awareness. After his flagship, *Honolulu*, fired on *Mikazuki*, the Japanese destroyer withdrew to the north. Ainsworth felt he had the enemy on the run and, at 0126, ordered McInerney to pursue. McInerney took *Nicholas*, USS *O'Bannon* (DD 450), and USS *Taylor* (DD 468) north while USS *Radford* (DD 446) and USS *Jenkins* (DD 447) stood by *Leander*. Ainsworth lost track of where they were.[50]

When new radar contacts appeared 23,000 yards off the port bow at 0155, Rear Admiral Ainsworth hesitated. He was "practically certain" that they were not his van destroyers, but he was not sure. Hastily exchanged radio messages failed to resolve the issue. At 0205, with the range rapidly closing, Ainsworth fired star shells to illuminate the four unidentified ships. As soon as the star shells burst, he ordered his cruisers to open fire, but it was too late. Shimai had completed reloading and made a perfect torpedo attack. *Honolulu* and *St. Louis* were both hit, along with USS *Gwin* (DD 433). The cruisers would survive with damaged bows, but *Gwin* had to be scuttled. *Leander* did not return to service until after the war.[51]

A Royal Navy observer felt that Ainsworth's failure to trust his CIC organization had kept him from "immediately" identifying the "second" enemy group as hostile. Had the CIC team been "unhampered by interference from the Admiral," the observer wrote, it "should have been able to keep track of the van destroyers" and distinguish them from the approaching enemy force. Unfortunately, Ainsworth believed that night actions typically developed "too quickly" for the CIC to operate as a "middle man."[52] He diagnosed the problem correctly—night actions *did* move quickly—but he failed to recognize that the solution was to distribute information processing to a skilled team that could rapidly make sense of all available information and present it in actionable form to the formation commander. Had he done so, the results at Kolombangara might have been quite different.

DESTROYER REVIVAL

After the battles of Kula Gulf and Kolombangara, the South Pacific Area was reduced to just one cruiser-destroyer force. Its new amphibious force commander, Rear Adm. Theodore S. Wilkinson, had commanded destroyers in the early interwar period and was familiar with their potential. He felt it would be more effective to let his destroyers operate independently to contest Japanese reinforcement efforts. On 5 August, after discovering that the Japanese were about to make another run to Kolombangara, Wilkinson ordered the new commander of Destroyer Division 12, Cdr. Frederick Moosbrugger, to sweep Vella Gulf in the hopes of intercepting the enemy ships.[53]

Moosbrugger had relieved Commander Burke just two days before. Burke was one of the Navy's best tacticians. He had studied prior battles, familiarized himself with Nimitz's revised instructions for Night Search and Attack, and developed a plan for using two destroyer divisions in concert to make stealthy attacks at night. In later years, Burke described his concept this way: "One division would slip in close, under cover of darkness, launch torpedoes, and duck back out. When the torpedoes hit, and the enemy started shooting at the retiring first division, the second half of the team would suddenly open up [with guns] from another direction. When the rattled enemy turned toward this new and unexpected attack, the first division would slam back in again."[54] Burke had trained his captains in these tactics and they wanted to use them; when Moosbrugger assumed command, they convinced him to adopt Burke's plan. Since Moosbrugger was familiar with destroyers and felt that night torpedo attacks were "second nature," their task was not difficult.[55]

Moosbrugger divided his six destroyers into two divisions. The first contained USS *Dunlap* (DD 384), USS *Craven* (DD 382), and USS *Maury* (DD 401). They had a larger torpedo armament and would attack first. The second, with USS *Lang* (DD 399), USS *Sterett* (DD 407), and USS *Stack* (DD 406), would hold back, ready to open fire with their guns. As they steamed up the west coast of Kolombangara late on 6 August, the night was very dark; there was no moon and "visibility [was] very low." At 2333, *Dunlap*'s SG radar made contact at 23,900 yards. Moosbrugger brought his first division into firing position, ordering a "full port torpedo broadside." Effective CIC plots allowed accurate torpedo firing solutions, and, at 2340, with the range down below five thousand yards, Moosbrugger launched his salvo at a column of four Japanese destroyers.[56]

In his action report, Moosbrugger described the next four minutes as "an eternity." Alert lookouts aboard *Hagikaze* and *Shigure* sighted Moosbrugger's ships just before the torpedoes arrived, but too late to avoid them. When the torpedoes hit, a series of "terrific explosions" shattered three of the Japanese ships. *Hagikaze* lost way and quickly came to a halt; *Arashi* started to burn;

MAP 6.2. BATTLE OF VELLA GULF, 6–7 AUGUST 1943

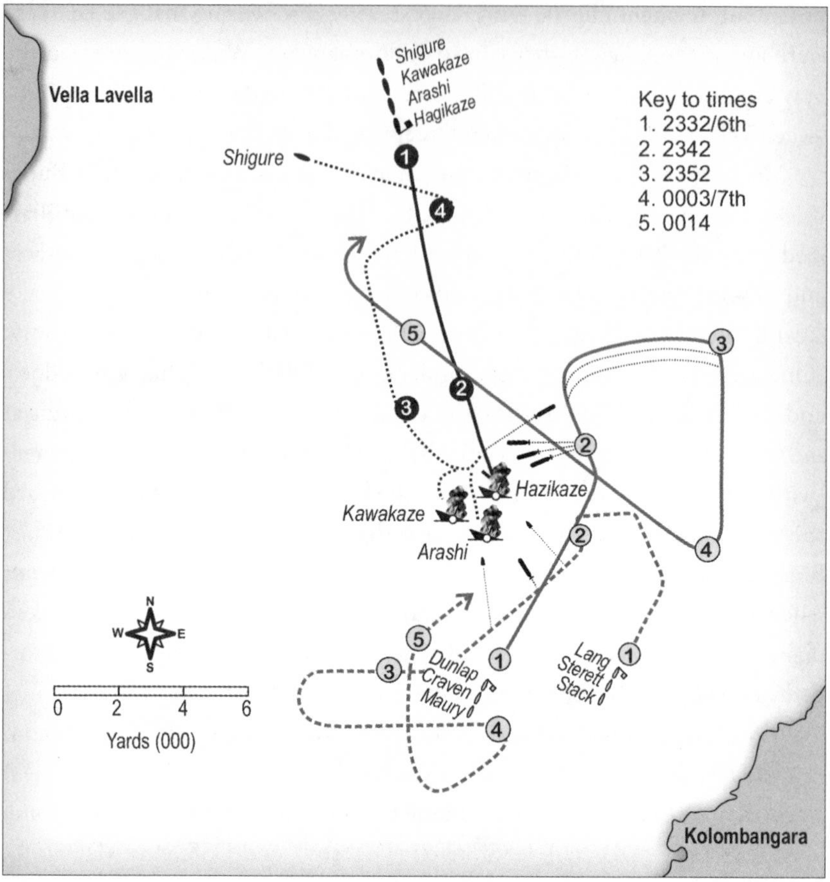

and *Kawakaze* "burst skyward in a ball of ruddy flame" as her forward maga-zine exploded. A dud hit *Shigure*'s rudder but failed to explode. Undamaged, she fired a salvo of torpedoes and tried to escape the "great mass of flames and explosions" by reversing course to the north.[57]

Moosbrugger's second division opened fire with its guns. His first division came back around and did the same. *Stack* launched four more torpedoes at the crippled Japanese destroyers. *Kawakaze* sank quickly. *Arashi* and *Hagikaze* attempted to return fire, but their aim was wild. Moosbrugger knew the battle was won and "it was only a matter of time before [the] complete destruction of the enemy." *Arashi*'s fires triggered a magazine explosion and the ship "virtually disintegrated." *Hagikaze* remained afloat, but six more torpedoes eventually finished her off. Only *Shigure* escaped.[58]

The Battle of Vella Gulf showed what the Navy's destroyers could do when allowed to capitalize on their strengths. The CIC permitted distributed formations to coordinate their movements, approach unseen, and make effective use of the destroyers' "most devastating weapon," the torpedo.[59] Although these concepts had been core to the Navy's initial plans for the Night Search and Attack, they had been abandoned in the interwar period. Wartime experience combined with new technologies and techniques to prompt a return to earlier principles. As Lt. Cdr. Francis T. Williamson, *Craven's* commanding officer, observed, "Our destroyer doctrine is sound, and our years of hard work will pay dividends, provided we use destroyers offensively and not for the protection of cruisers and battleships."[60]

In a series of battles that followed, Navy destroyers demonstrated the soundness of their doctrine and aggressively capitalized on the changing nature of the war in the South Pacific. Their increasing dominance was a gradual process. After Vella Gulf, Japanese formations became more cautious. They often accomplished their missions, but withdrew in the face of opposition, ceding control of the narrow waters of the Solomons to Wilkinson and his ships. In mid-August, when Halsey advanced to Vella Lavella, Wilkinson immediately blockaded Kolombangara to cut off the island and prevent a Japanese withdrawal. A brief engagement the night of 17 August failed to prevent the Japanese from establishing a base for their withdrawal effort, but it signaled their increasing desire to avoid decisive combat and preserve their ships.[61]

During the dark nights of late September and early October, the Japanese evacuated the bulk of their forces from Kolombangara. Wilkinson's destroyers sought to intervene. On the night of 27 September, five of them intercepted a group of Japanese barges, but attacks by enemy planes distracted them before they could sink very many. Two nights later, another group of Wilkinson's ships sank more barges and exchanged long-range fire with Japanese destroyers. Wilkinson had more success the night of 1 October; his destroyers sank submarine *I-20* and chased after a Japanese destroyer formation. The Japanese refused battle. They finished their evacuation the following night. Capt. William Cooke and his six destroyers tried to stop them. Cooke closed to within forty-five hundred yards of a Japanese destroyer formation and launched torpedoes before opening fire with his guns. Although a few shell hits were scored, Cooke was unable to inflict any significant damage.[62]

After evacuating Kolombangara, the Japanese attempted to remove their forces from Vella Lavella. The night of 6 October, Rear Admiral Ijuin Matsuji led an evacuation mission with six destroyers. Wilkinson committed two destroyer divisions—a total of six ships—to stop Ijuin. Capt. Frank Walker was ordered to take command of both divisions and "intercept and destroy [the] enemy evacuation force." The two divisions approached separately. Walker's USS *Selfridge* (DD 357), USS *Chevalier* (DD 451), and *O'Bannon* arrived at the rendezvous first, dogged by Japanese planes along the way. At 2231, when Walker established radar contact on two enemy groups at 19,500 yards, he did not hesitate. He set course for the enemy and increased speed. Instead of retiring, Ijuin resolved to fight. He hoped to bring his four ships— *Akigumo, Isokaze, Kazegumo,* and *Yūgumo*—together with *Shigure* and *Samidare* and "annihilate" the Americans. Walker moved quickly and disrupted Ijuin's plans. Intent on striking before the Japanese could concentrate, at 2245, Walker made visual contact. Ten minutes later, with the range down to just over seven thousand yards, he launched torpedoes at Ijuin's four destroyers and opened fire with his guns.[63]

Ijuin mistook Walker's ships for cruisers. As a result, he had estimated the range incorrectly and his destroyers—maneuvering in line abreast—were out of position for a torpedo attack. Only *Yūgumo* was clear to fire. She put eight torpedoes in the water and started shooting. Closest to Walker's ships, *Yūgumo* presented the best radar target. She was hit several times, caught fire, and slowed. However, at 2301, one of *Yūgumo*'s torpedoes slammed into *Chevalier*, detonating her forward magazine and tearing off her bow. *O'Bannon*, the next in line, could not maneuver around *Chevalier* in time and slammed into her crippled hull. *O'Bannon* was able to back away and take off some of *Chevalier*'s wounded, but Walker was down to his flagship *Selfridge*.[64]

At 2305, *Yūgumo* exploded. Walker turned to engage the other two Japanese destroyers. *Selfridge*'s guns straddled the *Shigure*, but one of the sixteen torpedoes that she and *Samidare* fired hit *Selfridge* just forward of her second gun mount at 2318. The explosion tore off her bow and knocked out all power, but fortunately no fires broke out. Despite crippling all three of Walker's ships, Ijuin retired. Japanese planes had warned him that more "cruisers"—the other American destroyer division—were on their way and rather than risk more losses, he withdrew.[65]

The Battle of Vella Lavella showed how the Navy had begun to dominate the waters of the Solomons by relentlessly hounding the Japanese with its destroyers. Although Ijuin had won a victory and accomplished his mission, he preserved his ships rather than finishing off Walker's force. Walker and other Navy destroyer commanders, aware that a campaign of attrition favored the United States because of its greater industrial capacity, were much more combative. After the battle, Walker was praised for his "skillful and aggressive" tactics. Other destroyer commanders would soon demonstrate similar aggressiveness and achieve greater success.[66]

WINNING THE NIGHT

The CIC was one of the reasons that Navy destroyer commanders became so aggressive. Recognizing its advantages, Walker "remained generally in [*Selfridge's*] CIC" during Vella Lavella.[67] Rear Adm. Mahlon S. Tisdale, the Pacific Fleet's type commander for destroyers, had issued a *CIC Handbook* in June 1943 that described how to maximize its value. Effective CICs integrated with the command function. By filtering information and eliminating extraneous details, they allowed captains and formation commanders to "concentrate on . . . decisions and carry the burden of command." At the same time, CICs helped ensure that orders were appropriately executed by coaching guns and torpedoes onto their targets. In situations where immediate action was vital—as with night aerial attacks—CICs were even starting to give orders to open fire.[68]

Increasingly capable CIC organizations "clarified the tactical situation," "facilitated quick decision making," and made it possible to coordinate distributed formations at night in ways that were previously impossible.[69] Burke's tactics—with two destroyer divisions acting in concert—were a beginning, but ultimately it became possible to fight in ways that reflected the Navy's tactics for daylight action, with destroyers closing for torpedo attack while larger ships fought from longer ranges in a battle line.

That ability became increasingly important as U.S. officers learned more about Japanese torpedoes. When Ainsworth fought Kula Gulf and Kolombangara, he assumed that Japanese torpedoes had capabilities broadly similar to Navy weapons. However, Japanese performance in those battles, along with new evidence, forced him to revise his assessment. In his comments on Kolombangara, Ainsworth noted that a 24-inch Japanese torpedo had "recently" been recovered off Guadalcanal. Its size suggested it would be "good for 14,000 yards at 32 knots" or "40 knots for 10,000 yards," the kind of performance that would have been required to hit *Helena* during Kula Gulf.[70]

Navy cruiser commanders began to change their tactics. Merrill, for example, assumed that Japanese navy torpedoes could reach 24,000 yards. He would counter that threat by fighting in three groups. Merrill planned to keep his cruisers "close to the maximum enemy torpedo range." Because it was nearly impossible to spot the 6-inch shell splashes of his cruisers' guns at those ranges, even with Mark 8 radars, Merrill held regular practices to familiarize his crews with the challenge of radar control at long range. Before the cruisers opened fire, Merrill planned to have his two destroyer divisions—one on either flank—make "an undetected approach to the torpedo firing position."[71]

When Halsey invaded Bougainville on 1 November 1943, Merrill had a chance to put his plan into action. Rear Admiral Ōmori Sentarō headed south from Rabaul with heavy cruisers *Myōkō* and *Haguro*, light cruisers *Sendai* and *Agano*, and six destroyers, intent on destroying the Allied transports in Empress Augusta Bay. Because the bulk of the Pacific Fleet was concentrating to the north for Operation Galvanic, the assault on the Gilbert Islands, Merrill's four cruisers "constituted the principal surface strength left in the South Pacific." He was intent on preserving them. Ōmori approached in three groups, with his heavy cruisers in the center and a light cruiser leading three destroyers on either flank. Merrill positioned his cruisers—*Montpelier, Cleveland*, USS *Columbia* (CL 56), and *Denver*—"across the entrance" to Empress Augusta Bay. Arleigh Burke, now a captain, commanded Merrill's destroyers. Burke personally led Destroyer Division 45, deployed to the north, and Cdr. Bernard L. Austin commanded Destroyer Division 46, to the south.[72]

The Battle of Empress Augusta Bay began at 0227 on 2 November, when contacts appeared on U.S. radar screens 38,000 yards away. As CICs plotted the movements of Ōmori's ships, his formation became clear. Merrill sent in Burke's van destroyers "almost immediately." They approached Ōmori's left flank column, headed by cruiser *Sendai*. At 0239, Merrill countermarched to keep his cruisers between Ōmori and the Bay. Soon thereafter, at 0245, Merrill ordered Austin to attack Ōmori's right flank. Burke's torpedoes were in the water by 0246 and hits were anticipated "six minutes later." However, after just three minutes, *Montpelier's* CIC informed Merrill that the Japanese had taken evasive action. Surprise had been lost. Merrill opened fire at 0249.[73]

MAP 6.3. BATTLE OF EMPRESS AUGUSTA BAY, 1–2 NOVEMBER 1943

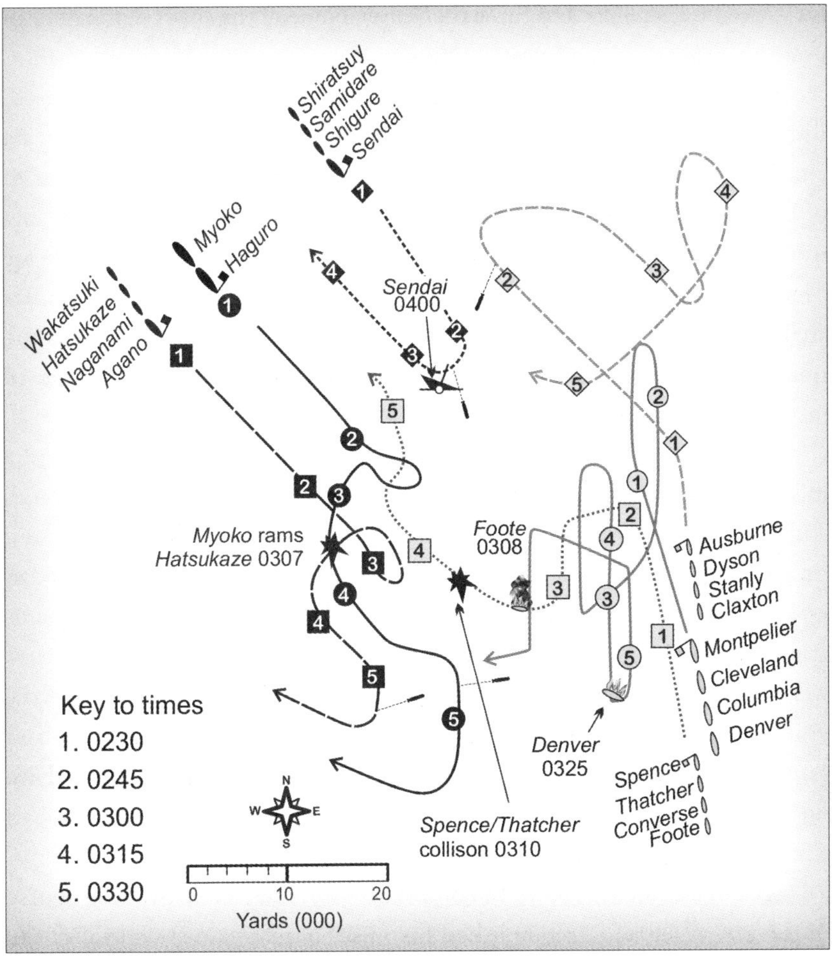

The range was about 20,000 yards. Rapid salvos from Merrill's cruisers began falling around *Sendai*. Hits jammed her rudder and started fires. The destroyers following her fired torpedoes, but two of them, *Samidare* and *Shiratsuyu*, collided as they maneuvered to avoid American shells. Ōmori's heavy cruisers "replied . . . almost immediately." Although their opening salvos were off, their "patterns were small," and when Japanese gunners had effective illumination "they were exceedingly accurate." Merrill said Japanese star shells were "extraordinary." To make them less useful, his cruisers fired star shells between the formations, forcing the Japanese to look through their glare. Merrill also made smoke and maneuvered radically to throw off Japanese aim.

These tactics were effective. Ōmori scored just five hits on Merrill's cruisers—two on *Columbia* and three on *Denver*. None of the shells detonated. In exchange, *Haguro* was hit several times.[74]

As the battle raged over a wide area, the situation became confused. Burke struggled to reform his division. Austin tried to consolidate his for an attack. USS *Foote* (DD 511) had gotten out of position during Merrill's reversal of course. At 0301, she was racing to join Austin when a Type 93—probably from *Samidare*—struck her stern. The crippled *Foote* became an obstacle for Merrill's cruisers and *Cleveland* nearly collided with her. Ōmori faced similar challenges. To avoid Merrill's gunfire, he took his heavy cruisers across the track of his right flank group. At 0307, *Myōkō* rammed *Hatsukaze* and *Haguro* just missed *Wakatsuki*.[75]

This gave Austin a great opportunity, but, at 0310, USS *Thatcher* (DD 514) misinterpreted his orders and sideswiped his flagship USS *Spence* (DD 512). Although the collision did little damage, it had serious consequences. Austin left the CIC and came to *Spence*'s bridge. The SG display there had been knocked out by the collision and, now "blinded," Austin lost track of the situation. Two decks below, *Spence*'s CIC concluded the Japanese heavy cruisers were "friendly" and a valuable opportunity to attack them while they were vulnerable was lost.[76] Instead, Austin's ships closed the crippled *Sendai* and sank her. Afterward, Burke mistook *Spence* for an enemy ship and opened fire on her before closing with *Hatsukaze* and finishing her off at 0539. Ōmori had decided to withdraw some two hours before.

Although Merrill's ability to control his forces degraded over the course of the action, he had accomplished his mission, maintained control of the approaches to Empress Augusta Bay, and won the battle. Merrill's improved tactics demonstrated how far the Navy had come over the past year. Rather than provoking a confused melee, his plan had exploited the strengths of his ships and allowed decentralized execution by his subordinates. Merrill had preserved his cruisers. None of his ships was lost. In his comments after the battle, Burke praised Merrill's "sound doctrine," "good battle plans," "aggressive spirit," and "faith in the ability of his subordinates."[77]

Later that month, on the night of 24 November, Burke would finally have an opportunity to demonstrate his skill as formation commander. The Japanese were making a "Tokyo Express" transport run to Buka. Halsey ordered Burke to take his Destroyer Squadron 23, "get athwart the Buka-Rabaul evacuation line," and intercept the enemy ships.[78] Burke had five destroyers divided into two divisions. He would lead Destroyer Division 45, with USS *Charles Ausburne* (DD 570), USS *Claxton* (DD 571), and USS *Dyson* (DD 572). Commander Austin led Destroyer Division 46, with USS *Converse* (DD 509) and *Spence.*

Reprising his tactics from earlier in the year, Burke planned to operate his divisions independently. Austin would provide cover while Burke's division approached to torpedo range. Once Burke's torpedoes arrived, the two divisions would attack from multiple directions and "rattle" the enemy. Burke lacked time for a conference with his captains, but they were all familiar with his tactics and doctrine. Radio messages were enough to allow them to act as a cohesive formation in the coming battle. As they headed northwest, Burke told them, "We will attack immediately on contact. We will not withdraw while a ship floats and can fight. We will not abandon a cripple."[79]

Burke's objective was a formation of five destroyers under Captain Kagawa Kiyoto. *Amagiri, Yūgiri,* and *Uzuki* were serving as transports, under the command of Captain Yamashiro Katsumori. *Ōnami* and *Makinami* operated as a screen under Kagawa's command. All of them arrived off Buka around 2300, 24 November, and, by 0045, were on their way back to Rabaul. As they made for their base, Burke reached his patrol position, slowed to twenty-three knots to keep his wakes hidden, and waited. The night was dark, with low clouds, no moon, and occasional rain. At 0141, Burke's radars detected Kagawa's two screening destroyers at a range of 22,000 yards. Burke quickly turned to intercept.[80]

By 0156, the range was down to forty-five hundred yards. Burke ordered his division to fire a half salvo of fifteen torpedoes at Kagawa's ships. After the torpedoes were in the water, Burke turned away and increased speed to thirty knots. A lookout aboard Kagawa's flagship *Ōnami* reported sighting the "shadows" of Burke's ships, but it was too late. Thirty seconds later, several torpedoes struck *Ōnami*. She disintegrated in "a ball of fire 300 feet high." *Makinami* was also hit, but remained afloat.[81]

Just before his torpedoes struck, Burke's radars picked up Yamashiro's transport destroyers, 13,000 yards astern of Kagawa's ships. Burke ordered Austin to finish off the cripples and set off after Yamashiro. The Japanese captain had other ideas. He turned north, intent on escaping to Rabaul. While Austin closed with and finished off *Makinami*, Burke engaged in a running battle with Yamashiro's destroyers. At 0212, "on a hunch," Burke changed course to the right briefly and avoided three torpedoes launched by *Yūgiri*. They detonated in *Charles Ausburne*'s wake; all of Burke's ships felt the resulting "heavy explosions."[82]

At 0222, with the range down to eight thousand yards, Burke opened fire with the forward guns of his three destroyers. When the enemy returned fire, Burke fishtailed his ships to bring their aft guns into action. With the range closing, the Japanese ships zigzagged and then split up, taking separate courses to get away. Burke pursued *Yūgiri*, the ship with the largest radar signature. She was hit repeatedly, and at 0305 a shell detonated in her machinery spaces, ensuring that she would not escape. After an "intensive bombardment," Burke's destroyers sank her at 0328.[83]

PHOTO 6.4. Capt. Arleigh A. Burke, commander of Destroyer Squadron 23, and the other victors of the Battle of Cape St. George enjoy a beer at "Cloob Des-Slot" at Purvis Bay near Tulagi on 24 May 1944. *From left,* they are Cdr. Roy A. Gano, commanding officer, USS *Dyson* (DD 572); Cdr. Luther K. Reynolds, commanding officer, USS *Charles Ausburne* (DD 570); Burke; Cdr. Bernard L. Austin, commander, Destroyer Division 46; Cdr. DeWitt C. E. Hamberger, commanding officer, USS *Converse* (DD 509); Cdr. Herald Stout, commanding officer, USS *Claxton* (DD 571); and Cdr. Henry J. Armstrong, commanding officer, USS *Spence* (DD 512). *U.S. Naval History and Heritage Command, NH 59864*

Burke tried to order Austin to pursue the other fleeing Japanese ships, but their two destroyer divisions were at the edge of TBS radio range. Although a test communication got through to Austin, Burke's orders did not. Once Austin rejoined, Burke set off in pursuit, hoping to catch at least one more Japanese destroyer, but at 0405, with dawn approaching and no enemy in sight, Burke elected to retire.[84]

Vice Adm. William S. Pye, president of the Naval War College, called Burke's victory at the Battle of Cape St. George a "classic." Pye correctly attributed "increasing American success" in the Solomons to "improvement in our organization," the "skill of our personnel," and the CIC, which conferred "advantages not dreamed of a short time ago."[85] Burke's assessment was similar. He praised his CICs and said they "performed admirably," preventing any "confusion" or "lack of knowledge." Remarkably, after the battle, "the track charts [from different ships] . . . superimposed one over the other."[86] The shared view of the battle created by their CICs enabled Burke's captains to act as a unit and maintain cohesion throughout the action.

THE BATTLE OF SURIGAO STRAIT

During 1944, the Navy continued to improve CIC practices and enhance night-combat capabilities. In October, the dual offensives of Gen. Douglas MacArthur and Nimitz combined during the invasion of Leyte in the Philippine Archipelago. The assault triggered the Japanese *Shō-1* plan, an attempt to win a decisive battle on land, at sea, and in the air. For the Japanese navy, Shō-1 called for a two-pronged attack on the amphibious forces in Leyte Gulf. Halsey's Third Fleet concentrated against one of those prongs, the First Striking Force, led by Vice Adm. Kurita Takeo. Kinkaid, commander of the Seventh Fleet, ordered Rear Adm. Jesse B. Oldendorf to combine his Fire Support Group with the Close Covering Group led by Rear Adm. Russell S. Berkey and destroy the second.

Oldendorf worked out detailed plans in conference with Berkey and other subordinates. Although they had not operated together before—Oldendorf's ships fought in the Central Pacific and Berkey's in the Southwest Pacific—new doctrinal material allowed them to quickly agree on an approach and succinctly communicate it to their subordinates. *USF-10A, Current Tactical Orders and Doctrine, U.S. Fleet,* had been issued in February 1944. It built on an earlier document, *PAC-10, Current Tactical Orders and Doctrine, U.S. Pacific Fleet,* issued by

Nimitz in June 1943. These new manuals provided a framework that mitigated the problems associated with throwing task forces together without adequate preparation. *PAC-10* and *USF-10A* were designed to allow "forces composed of diverse types, and indoctrinated under different task force commanders, to join at sea on short notice for concerted action against the enemy."[87]

Oldendorf used a plan from *USF-10A* and positioned his forces to block the northern exit to Surigao Strait. He formed a battle line with six old battleships—USS *West Virginia* (BB 48), USS *Maryland* (BB 46), USS *Mississippi* (BB 41), USS *Tennessee* (BB 43), USS *California* (BB 44), and USS *Pennsylvania* (BB 38)—and placed cruisers and destroyers on both flanks. The left (east) flank had five cruisers, including Oldendorf's flagship USS *Louisville* (CA 28) and the nine ships of Destroyer Squadron 56, led by Capt. Roland N. Smoot. Three cruisers were on the right flank, along with the six ships of Destroyer Squadron 24, commanded by Capt. Kenmore M. McManes. Oldendorf deliberately made his left flank stronger to guard against enemy ships slipping around the east of Hibuson Island.[88]

Because Oldendorf's battleships had relatively few armor-piercing shells, he planned to fight at moderate ranges and employ guns and torpedoes together. His tactics echo prewar concepts that envisioned destroyers making close-range torpedo attacks under the cover of battleship gunfire. However, Oldendorf's destroyers would use their radar, CICs, and lessons from the fighting in the Solomons to maximize their striking power. They ambushed the first Japanese force to enter the strait and significantly reduced its strength long before the battleships engaged.[89]

The first destroyer attacks were made by five pickets from Task Group 79.11. Their commander, Capt. Jesse G. Coward, a veteran of the First Battle of Guadalcanal, joined Oldendorf on his own initiative, and, when reports from PT boats in the southern part of the strait signaled the approach of the Japanese, Coward set off to find them. His flagship USS *Remey* (DD 688) led USS *McGowan* (DD 678) and USS *Melvin* (DD 680) down the eastern side of the strait, while Cdr. Richard H. Phillips took USS *McDermut* (DD 677) and USS *Monssen* (DD 798) down the western side. At 0240, *McGowan* reported a radar contact at 37,600 yards almost due south. In time, it resolved into at least three ships, heading north.[90]

McGowan's CIC was tracking the Third Section (First Striking Force) led by Vice Admiral Nishimura Shōji, with battleships *Yamashiro* and *Fusō*, cruiser *Mogami*, and destroyers *Shigure, Michishio, Asagumo*, and *Yamagumo*. Nishimura was maneuvering into his battle formation, a single line with the destroyers in the van, battleships in the center, and *Mogami* in the rear. Behind Nishimura was the Second Striking Force, under Vice Admiral Shima Kiyohide. Nishimura and Shima did little to coordinate their movements because they apparently had very different objectives. Nishimura's mission was to sacrifice his battleships to help ensure Kurita's penetration of Leyte Gulf. Shima hoped to support this effort, but planned to withdraw if the odds were against him.[91]

As Coward closed for a torpedo attack, lookouts on board *Shigure* sighted his ships. Japanese searchlights swept the darkness, but Coward's destroyers were beyond their range. Nishimura apparently assumed it was a false contact; he took no evasive maneuvers as Coward closed. Coward thought he had been detected. Although Coward and Phillips planned to make a coordinated attack and strike from both bows of the enemy simultaneously, they had lost contact with each other. Determined to attack while he could, at 0300, Coward fired torpedoes from long range, 11,500 yards. As he did, flash eliminators on his flagship, *Remey* failed, and a searchlight fixed on her. The Japanese opened fire, but Coward made smoke as he turned away, and no hits were scored.[92]

Nishimura continued to hold his course. Most of Coward's torpedoes missed, but *Melvin* scored two hits on *Fusō*. A "tremendous explosion" erupted on *Fusō's* starboard side, and as she sheered out of line a fire broke out.[93] Just before, at about 0307, Phillips' destroyers appeared in the darkness ahead and the Japanese opened fire. Phillips kept his guns silent, hoping to present a difficult target and keep his ships obscured. At 0310, they began firing torpedoes from about nine thousand yards. This time Nishimura changed course, but his maneuvers put his destroyers in the path of the oncoming torpedoes. At 0319 *Yamagumo* exploded and sank. *Michishio* was crippled and brought to a stop. *Asagumo* lost her bow, but was able to retire down the Strait. Another torpedo hit *Yamashiro*, slowing her briefly, but she regained speed. *Fusō* fell behind and sank sometime around 0345. Coward's five picket destroyers had executed a prototypical Night Search and Attack and cut Nishimura's strength in half.[94]

Because Coward's destroyers encountered little resistance, Berkey released the right flank destroyers early. McManes maintained an "excellent picture" of the developing battle from the CIC of his flagship USS *Hutchins* (DD 476).

He divided his squadron into two attack groups; both came at Nishimura from the western side of the strait. While McManes took *Hutchins*, USS *Daly* (DD 519), and USS *Bache* (DD 470) to the south to cut off a potential Japanese retreat, his trailing group of HMAS *Arunta*, USS *Killen* (DD 593), and USS *Beale* (DD 471) closed and fired torpedoes at 0323 from a range of about eight thousand yards. *Arunta* fired a broader spread of torpedoes at a higher speed than her American counterparts—an important difference in doctrine that allowed for more errors in her firing solution. Most of these torpedoes missed, but at 0331, one of them, probably fired by *Killen*, hit *Yamashiro*.[95]

By 0330, McManes had reached his attack position, and *Hutchins* commenced firing torpedoes from eighty-two hundred yards. *Daly* and *Bache* held theirs until 0335. None of these torpedoes hit because by this time Nishimura's formation had disintegrated and it was difficult to develop clear firing solutions. McManes opened fire with his guns at 0341 from a range of 12,000 yards and "heavily hit" *Michishio*. *Yamashiro* fired back, placing shells in *Hutchins'* wake, but none scored. Although ordered to retire, McManes closed *Asagumo* and launched another torpedo spread; these missed their intended target but found the drifting *Michishio* and sank her. On McManes' way north, his destroyers continued to fire on *Yamashiro* and *Mogami*.[96]

At 0335, as Nishimura's remaining ships approached Oldendorf's battle line, he released Smoot's Destroyer Squadron 56. Smoot divided his squadron into three attack groups: his flagship, USS *Newcomb* (DD 586), with USS *Richard P. Leary* (DD 664) and USS *Albert W. Grant* (DD 649), would attack down the center of the strait while USS *Robinson* (DD 562), USS *Halford* (DD 480), and USS *Bryant* (DD 665) attacked from the northeast and USS *Heywood L. Edwards* (DD 663), USS *Leutze* (DD 481), and USS *Bennion* (DD 662) struck from the northwest. As they approached their attack positions, a signal from Capt. Thomas F. Conley, commander of Smoot's eastern attack group, captured the moment: "This has to be quick. Stand by your fish."[97]

Before any of Smoot's ships could launch their fish, Oldendorf opened fire. The cruisers on both flanks started shooting at 0351 and the battleships at 0353. The range for the battleships was about 23,000 yards. At 0356, Smoot's eastern attack group launched torpedoes from around nine thousand yards. The confused radar picture—shell splashes were landing around the Japanese ships—and *Yamashiro's* turn to port caused them all to miss. A minute later, Smoot's western attack group launched from eight thousand yards. Soon after,

MAP 6.4. BATTLE OF SURIGAO STRAIT, 24–25 OCTOBER 1944

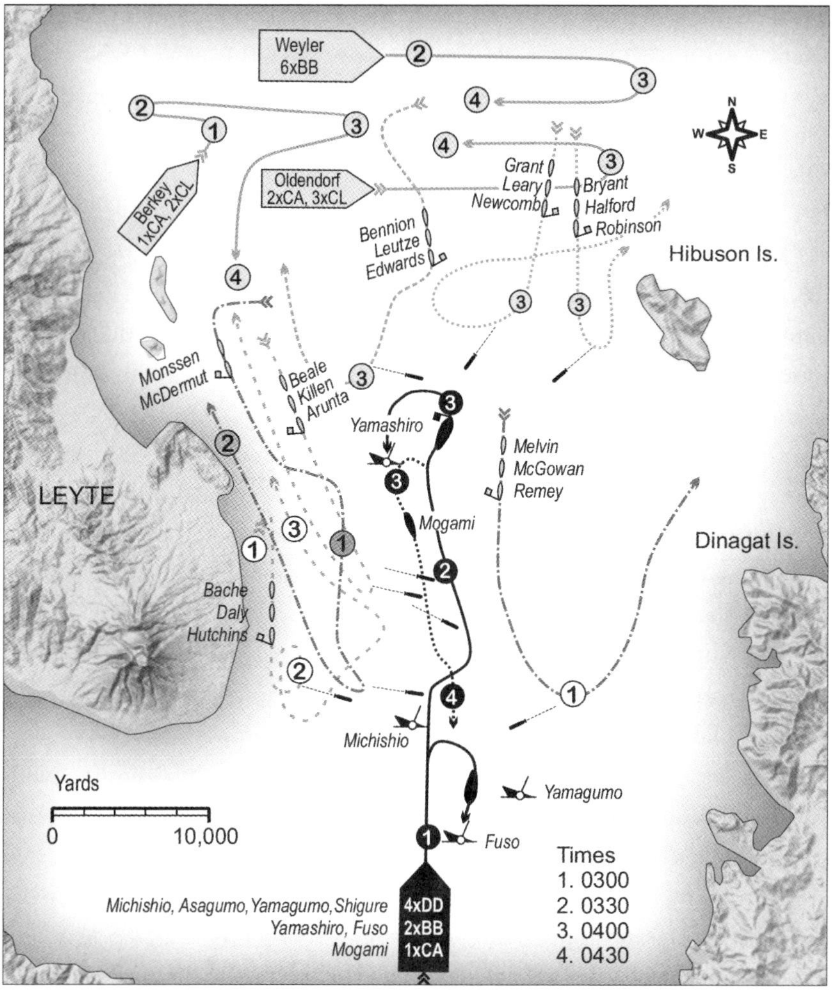

Bennion sighted a second target and launched another five torpedoes at it. One of these found *Yamashiro* at about 0405. Smoot pressed his own attack home, firing at 0404 from just sixty-two hundred yards. *Albert W. Grant* was straddled by ships from both sides and hit repeatedly. Her crew quickly launched her remaining torpedoes while they could still be used. Japanese torpedoes streamed by but missed. *Yamashiro*, the old Japanese battleship, was not so lucky; she was hit again around 0411.[98]

Yamashiro had taken tremendous punishment throughout the battle. *West Virginia*, the fleet's former gunnery champion, was the first of Oldendorf's battleships to open fire. Using Mark 8 radar, *West Virginia's* fire-control team had a solution in place at 30,000 yards. Her first salvo hit and "crushed the enemy's great pagoda tower like a sand castle."[99] All thirteen of *West Virginia's* salvos straddled the target. *Tennessee* and *California*, also equipped with the Mark 8, started firing accurate salvos soon after *West Virginia*. *Maryland*, *Mississippi*, and *Pennsylvania* were still using the Mark 3; they had trouble developing clear solutions, and fired infrequently, if at all.[100]

After the last torpedo hit, *Yamashiro* sank within five minutes.[101] Nishimura's attempt to penetrate Leyte Gulf failed. Shima would not be so bold. After entering the battle area, he fired torpedoes at a radar target to the north—most likely Hibuson Island—and then withdrew. His decision to retire was hastened by the fact that his flagship, *Nachi*, having incorrectly estimated *Mogami's* course and speed, collided with the damaged cruiser. Shima would manage to escape. *Mogami* did not; Oldendorf's cruisers and destroyers caught her and *Asagumo* the next morning. They sank *Asagumo*. *Mogami* remained afloat long enough to be bombed by American planes at 0915. Crippled and drifting, she was scuttled by a torpedo from destroyer *Akebono*.[102]

The Battle of Surigao Strait was a complete victory. The CIC, modern fire-control radars, more cohesive formations, and improved tactical doctrines allowed Oldendorf to combine the most effective prewar concepts—coordinated destroyer torpedo attacks in concert with battleship gunfire—with wartime lessons. Even though Surigao Strait was a night battle, Oldendorf and his subordinates used the increasing capabilities of the Navy's ships and doctrines to fight it much like a daylight action, coordinating their ships in the darkness in ways incomprehensible just a few years before.

However, the battle for Leyte continued long after the success at Surigao. U.S. Navy destroyers regularly conducted interdiction missions to disrupt Japanese resupply and reinforcement efforts. Not all these missions were successful. On the night of 2 December 1944, Cdr. John C. Zahm took destroyers USS *Allen M. Sumner* (DD 692), USS *Moale* (DD 693), and USS *Cooper* (DD 695) into Ormoc Bay on the island's northwestern edge. Enemy planes started attacking at 2308 and for the next two and a half hours, Zahm was in a running battle. The Japanese had learned that isolated attacks from multiple directions could disrupt the air defense systems of the Navy's carrier task forces; they took a similar approach with Zahm.[103] A new aerial attack developed, on average, "every eight and one half minutes."[104] The incessant attacks overwhelmed

CIC personnel and their communication circuits, keeping Zahm from developing a clear picture.[105]

Once inside the bay, Zahm identified two surface targets and opened fire at 0005. These were the small Japanese destroyers *Take* and *Kuwa*, advancing to protect the transports behind them. When Zahm opened fire, *Take* and *Kuwa* responded, as did shore batteries and automatic cannons on the transports. The Japanese had the advantage of being obscured against the shore while Zahm's destroyers were visible in the moonlight. Radar-controlled fire sank *Kuwa* and damaged *Take*, but she fired a torpedo that hit *Cooper* at 0013 and broke her in two. The U.S. destroyer sank in less than three minutes. Zahm was unaware that he had lost her until 0029. By that point, having disengaged to "clarify the situation," he elected to withdraw.[106] The relatively unharmed Japanese transports departed the next day, escorted by the damaged *Take*. *Cooper's* loss—the last of the Navy's major combatants lost in surface action with the Japanese—demonstrated the risk and uncertainty that dominated night combat, even in the face of revolutionary new methods.

CONCLUSION

The U.S. Navy's experience with night combat in World War II demonstrates the challenge of integrating new technologies into tactics and doctrine. In the confused battles off Guadalcanal in 1942, Navy ships had radars, but the information that those radars provided was not used effectively. It took new practices, procedures, and organizational structures—the CIC and its information-processing mechanisms—to capitalize on radar's full potential. Once Navy officers began to recognize what the CIC could do and adjusted their behavior accordingly, the Navy's approach to night combat was transformed.

However, the path that transformation took was not wholly new. Wartime experience demonstrated the wisdom of earlier tactics, those that emphasized stealth, surprise, and coordinated action by small groups of destroyers. Nimitz stressed these ideas in his revised instructions for Night Search and Attack in November 1942. Burke and others developed effective means for employing them. Once Wilkinson assumed command of the South Pacific Amphibious Force, his faith in the ability of destroyers to operate on their own combined with increasing cruiser losses—and growing recognition of the risks of operating cruisers in the narrow waters of the Solomons—to give destroyer commanders their chance. Moosbrugger successfully employed Burke's tactics at the Battle of Vella Gulf in August 1943 and introduced a new paradigm, one that combined the revolutionary potential of radar with aggressive tactics based on stealth and surprise.

With the CIC, the Navy could capitalize on doctrinal concepts that it had emphasized for decades. Destroyers could close for torpedo attacks; cruisers and battleships could remain at longer ranges and use their guns. Even at night, different ship types could operate in distributed formations and coordinate their actions because the CIC provided ship and formation commanders with a broadly similar view of the battle. Empress Augusta Bay and Surigao Strait demonstrated the revolutionary potential of the Navy's increasing capabilities, and both were highlighted as examples of "sound tactical procedures" in postwar manuals.[107] However, even in these exemplary battles, the situation gradually became confused, as smoke, rain squalls, and shell splashes obscured enemy ships and hid friendly vessels from sight. Radars and CICs could not transform night into day, but they did provide the Navy's ships and their commanders with much greater situational awareness, enabling them to master the former masters—their Japanese opponents—in night combat.

CHAPTER 7

CONTROLLING THE CHOPS
DESTROYER NIGHT ACTION AND THE BATTLE OF ILE DE BATZ, OCTOBER 1943–JUNE 1944[1]

MICHAEL WHITBY

"There you are; give him the works." So said Commander Harry DeWolf, Royal Canadian Navy, to his gunnery officer on the destroyer HMCS *Haida* on the night of 26 April 1944.[2] With that, *Haida*'s 4.7-inch main armament began raking the stopped and burning German torpedo boat *T-29* in the English Channel off northern Brittany. Three other Tribal-class destroyers of the Royal Navy's Tenth Destroyer Flotilla (10 DF) joined in, and tracers stitched the night as they hammered the enemy for forty-five minutes. The crossfire threatened friends; DeWolf later found a shell fragment with British markings embedded in his golf clubs. When guns failed to dispatch *T-29*, the Allied destroyers tried torpedoes, but all missed. Finally, after more pounding, *Haida*'s logbook rejoiced: "ENEMY HAS SUNK!!"[3] With celebration came palpable relief. Britain's Plymouth Command, to which the Canadians were assigned, had sought such an outcome since the autumn of 1943, but achieving success at night had proven elusive. Once gained, though, the 10 DF proved as capable at that warfare as similar units in World War II, and the process through which they attained that proficiency provides lessons of enduring relevance.

BACKDROP

With Germany's occupation of French bases in 1940, the western entrance to the English Channel—the "Chops of the Channel"—where the Narrow Seas meet the Atlantic Ocean—became the setting of incessant sparring between the light forces of the Kriegsmarine and Britain's Royal Navy. In the spring of 1944, however, the seas between the major British base at Plymouth and the coast of northern Brittany assumed heightened significance. Planners for Operation Neptune, the invasion of northwest France, realized that control of the western Channel was vital to success. Likewise, the Kriegsmarine knew it had to interdict Allied shipping to have any chance of thwarting Neptune, and since the bulk of its destroyer strength was based on the Biscay coast it had to breach the Chops to attack the invasion. The predominance of air power meant that both the Allies and the Germans chose to conduct these operations under the cloak of darkness.

This chapter outlines how the Royal Navy and the Plymouth Command grappled with the art of night fighting. This occurred as growing faith in radar, or radio direction-finding (RDF), as the British initially called it, sparked debate on how the evolving technology influenced traditional tactics—in particular, the goal of seizing the initiative at the outset of battle, a central element of the Royal Navy's night-fighting doctrine. Radar had an enormous impact on how that maxim could be achieved; and the Royal Navy's endeavors to take best tactical advantage of that, along with the Kriegsmarine's response, are major components of the story ahead. Ship design, weaponry, intelligence, and force development also were factors, and since it was people who did the actual fighting, leadership, character, and experience played a role, including within the Kriegsmarine destroyer force, whose professionalism often is overlooked. In the end, Neptune was an Allied victory; Canadian destroyers played an instrumental role in the fighting and became a celebrated element of Canada's naval heritage.

Early in the war, naval combat in the Chops largely involved scraps among light forces, with British units attempting to parry Kriegsmarine attacks on coastal shipping. Destroyer vs. destroyer fighting was rare; however, an action in the predawn hours of 29 November 1940, when five ships of the Royal Navy's Fifth Destroyer Flotilla (5 DF) confronted three German counterparts, illuminated the challenges that could arise, and provides a benchmark of how the navies concerned progressed in their night-fighting capability.

That night the 5 DF, under Captain Lord Louis Mountbatten, patrolled southwest of Plymouth. With his own ship under repair, Mountbatten embarked in HMS *Javelin*, captained by Commander Anthony Pugsley, accompanied by HMS *Jupiter*, HMS *Kashmir*, HMS *Jackal*, and HMS *Jersey*,

MAP 7.1. THEATER OF OPERATIONS

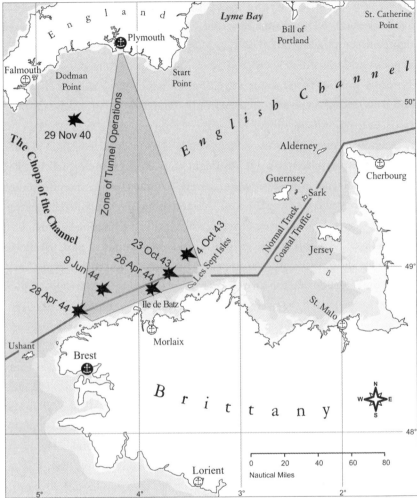

modern destroyers commanded by seasoned professionals. Line-ahead, where ships "followed Father," was the Royal Navy's standard night formation, but Mountbatten thought his captains would have difficulty remaining closed up at the high speed of twenty-seven to thirty knots that he intended to employ, so he utilized a line-of-bearing formation, where ships took station offset from the leader.

Confirming reports that enemy destroyers were about, at 0521 Mountbatten sighted a searchlight bearing 340 degrees fine on the starboard bow. He turned toward; then, at 0537, "a turn to 310° was made, to make sure of keeping between the enemy and his base." From *Javelin's* bridge, a dozen pairs of eyes probed the darkness—the destroyer's temperamental Type 286 radar had packed up.[4] Mountbatten reported: "At 0540 two darkened vessels were sighted fine on the starboard bow, crossing ahead of our line from starboard to port, and I immediately ordered 'Alarm Bearing 310' to be passed by W/T. The range was estimated at 5,000 to 6,000 yards and was closing rapidly." Fortuitously, *Javelin's* Type 286 came back online and detected three ships at 3,500 yards, near-point-blank range.[5]

Pugsley recalled a hurried exchange on *Javelin's* bridge. He wanted to initiate; Mountbatten chose to react.[6] "Straight on at 'em, I presume, Sir?" Pugsley suggested. "No, no," Mountbatten demurred. "We must turn to a parallel course at once or they will get away from us." Mountbatten later explained he wanted "to try to get as close to the enemy as possible, whilst remaining between him and his Base."[7] Accordingly, Pugsley brought *Javelin* around to port, parallel to the enemy, and gave the order to open fire. The director control tower lost bearing on the target during the turn, and a minute passed before *Javelin's* 4.7-inch guns let loose. Due to the overestimate on range, her first rounds were "overs." Before the gunners could apply corrections, two torpedoes exploded against *Javelin's* starboard side, bringing the destroyer to a dead stop.

The friction that so often assails night action settled on the British formation. *Jupiter* swung out on *Javelin's* engaged quarter to maintain the line-of-bearing, but then had to veer sharply to port to avoid collision when the torpedoes hit *Javelin*; the explosions temporarily blinded those on *Jupiter's* bridge. *Kashmir*, the third ship in formation, under Commander Henry A. King, the flotilla's second in command, was showered with oil from *Javelin*, and her bridge team's vision was impaired by her own gun flashes. Five minutes passed before King realized that *Javelin* was *hors de combat*. At that point, King led the remaining destroyers south at twenty-eight knots to get ahead of the enemy vessels, which had disappeared into the darkness. With nothing in sight after two hours, King doubled back north hoping to meet the enemy, but the Germans had won the race south. The Chops were empty. In *Javelin*, wallowing with her bow and stern blown off, Mountbatten had watched forlornly as his ships disappeared into the night. Sturdy *Javelin* survived, but forty-six crew members died.[8]

Beyond the German crews' skill and comfort at night, the Kriegsmarine's performance illustrated its doctrine. Operating under the rubric of the strategic defensive, destroyer commanders were directed to engage only targets of opportunity and to withdraw if they encountered serious opposition—"shoot-and-scoot," as one British officer derisively dubbed the doctrine during World War I.[9] This should not be construed as a lack of boldness; the Kriegsmarine simply lacked the strength to risk destroyers as expendable assets. Moreover, as one British officer posited, a force bent on evasion has an advantage at night, which on this occasion proved accurate.[10]

German destroyers *Karl Galster*, *Hans Lody*, and *Richard Bietzen* had begun their return run to Brest after sinking two small trawlers when their radar, an early model of FuMO-24, detected contacts, confirmed by the sighting of "five or six destroyers." Wheeling southwest to open from the enemy, the Germans fired torpedoes as they turned. Given the high speed and relative point-blank range of their targets—about fifteen hundred yards—there was only time to snap off a quick pattern. Nonetheless, two torpedoes found *Javelin*. Straddled by British shellfire, the three destroyers escaped south at thirty-five knots, suffering no casualties and only minor splinter damage. The commander of Marine Gruppe West considered the action "a complete success."[11]

While German sailors celebrated, there was tooth-sucking in Britain. At the Admiralty, which was responsible for identifying and disseminating operational lessons, criticism focused on the delay in getting into action. Citing doctrine, one officer thought that Mountbatten had displayed poor judgment:

> It would appear that the instructions in paragraph 8 of the Fighting Instructions were not sufficiently borne in mind. Here it is stated: "In any action at night, the primary object is to develop the maximum volume of gun and torpedo fire before the enemy can do so, and all other considerations are of secondary importance. Results at night will depend on the action taken in the first minute or so. . . . " A good deal of time seems to have been wasted in maneuvering for position and in disposing the destroyers on the correct line of bearing, instead of going straight for the enemy and engaging with all weapons.

Another agreed: "The outstanding feature is that we apparently had the advantage of first sighting but held our fire for at least three minutes. (0540–0543)." Summing it up, the vice chief of naval staff stressed that it was "elementary that one should open fire first at night."[12]

Additional comments showed that night-fighting procedures were in a state of flux, with debate over subjects that previously had been considered fundamentals. The Commander Western Approaches thought Mountbatten should have used searchlights to maintain contact with the enemy; the Admiralty disagreed, an officer noting, "A searchlight badly used or outside effective range is a known and proven danger." That view held, and searchlights, the primary visual night-fighting aid in World War I, were not used in any of the actions discussed in this chapter. Staff analysts highlighted the importance of radar—the device that made searchlights redundant—but lamented its poor reliability and limited availability. There also was criticism of the confusion caused by "fiddling about on lines of bearing at night." Finally, there was agreement the 5 DF's night-fighting efficiency was substandard, with more specialized training required.[13] As later events demonstrated, senior Royal Navy officers remained far from agreement on the essentials of night fighting, and shortcomings from November 1940 endured. That said, even with the march of technology, the principle of hitting first—however achieved—remained sacrosanct.

INNOVATION

Following the November 1940 victory, the Kriegsmarine shifted its thin destroyer resources to Norwegian and Baltic waters, leaving night operations in the Narrow Seas to their *Schnellboot* and other light forces. Reinforcements came in the summer of 1943, when the Germans deployed new Type 39 *Flottentorpedoboot*, or Fleet torpedo boats, to Brest to escort convoys running along the coast of north Brittany. Displacing 1,297 tons, capable of thirty-three knots, and armed with four 4.1-inch guns and six 21.7-inch torpedo tubes, the small destroyers—"Elbings" to the British—proved a thorn in the side of Allied naval forces. That said, they lagged their Allied counterparts in the critical area of radar.[14]

Although Germany had been at the cutting edge in the development of naval radar—the pocket battleship *Graf Spee* took a set to sea in 1936—chronic interservice rivalry had stifled development. The Luftwaffe claimed priority over equipment, forcing the Kriegsmarine to rely upon hand-me-downs that were ill-suited to naval warfare. Even late in the war, destroyers retained 1940-vintage FuMO24–FuMO26 series metric sets that were limited in range and discrimination. Less-effective A-scopes, where echoes formed vertical spikes on a horizontal range scale, were used for display, and the location of the aerial at the base of the foremast caused a 30-degree blind-spot astern. Kriegsmarine action reports from 1943 through 1944 reveal that although radar gave accurate ranges it had limited search capability.[15]

By contrast, the Royal Navy maintained steady progress. At the time of the November 1940 action, understanding of radar among most operational and staff officers was about as rudimentary as *Javelin's* Type 286. Within two years, however, "attitudes had changed dramatically. Radar was not just another burden, it was essential equipment."[16] By mid-1943 most British Fleet destroyers had two or three sets: gunnery, warning combined (for air and surface search), and warning surface. For gunnery, variants of Type 285 had evolved into an effective fire-control set. Besides providing continuous ranges it could detect shell splashes of "overs" and "unders" and, to the discomfort of operators, incoming rounds. The warning combined set used by destroyers was Type 291, a pioneering metric set that could detect destroyers out to nine miles, but by 1943, since radar was easily monitored or jammed, it was not switched on until action was joined. The most effective search radars were the warning surface sets. At mid-war most destroyers featured a variant of Type 271, whose narrow 10-cm beam produced superior range, discrimination, and accuracy. Under optimal conditions it could detect destroyers at about nine miles—essentially out to the horizon. Typical of the early days of radar, however, the equipment had its weaknesses. In early models the aerial was rotated manually, and after each sweep the operator had to reverse direction so the cord would not become tangled. Type 271 aerials also had to be located next to their power source, which was too heavy for destroyers' tripod foremasts. Nor could they be mounted atop the bridge superstructure without removing essential fire-control equipment; instead, sets were placed midships where searchlights once had been. The comparatively low height of this arrangement reduced the range of the radars; worse, destroyers' forward superstructure "wooded" the beam, causing a significant blind-spot ahead, as broad as 54 degrees. Type 285 or 291 could cover this gap in a pinch or ships could zigzag to open radar arcs, but the blind-spot still brought consequences.[17]

Since radar was relatively new, using fragile vacuum-tube technology, teething troubles were inevitable. Sets took time to warm up; breakdowns were common, especially under pounding from hard steaming or shock from

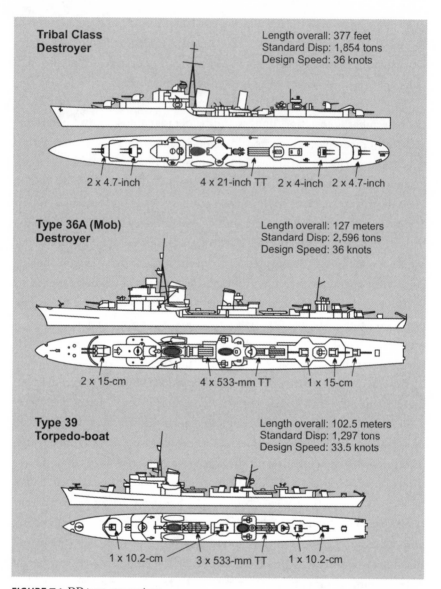

FIGURE 7.1. DD type comparison

main armament blast; and climatic conditions often impaired performance. Skilled operators and maintainers were in short supply. Nonetheless, by midwar the Royal Navy's radar advantage over the Kriegsmarine was such that one historian avowed that German sets "were to the Allied sets as a pocket torch is to a car headlight."[18] Even so, the tactics of how to take advantage of that superiority remained to be ironed out.

CALAMITY

In September 1943 the Plymouth Command initiated a series of offensive sweeps along the north coast of Brittany under the generic designation of Operation Tunnel.[19] Enemy destroyers and merchant shipping were both objectives, but the principal mission remained paring down Kriegsmarine destroyer strength. "It has always been hoped by all, including the forces concerned, that enemy destroyers w[ou]ld be the prize," one senior officer confessed.[20] From the start Vice Admiral Ralph Leatham, the commander in chief of Plymouth Command, was hamstrung by the limited forces at his disposal. Light cruisers and Fleet destroyers were best suited to the mission, but none was permanently attached to his command, so he had to rely on ships temporarily detached from the Home or Mediterranean fleets. The first four Tunnels run in September featured the cruiser HMS *Phoebe* and the Fleet destroyers HMS *Grenville* and HMS *Ulster*. They never met the enemy, but they had begun to evolve into a cohesive unit when *Phoebe* suddenly was ordered to the Mediterranean. With no other cruisers on hand for the next Tunnels run in October, Leatham reinforced *Grenville* and *Ulster* with smaller Hunt-class destroyers. As two actions show, deploying mixed forces unfamiliar with one another—compounded by the Hunts' lower speed—proved troublesome.

The first action came on the night of 3–4 October 1943, when the Hunts HMS *Limbourne*, HMS *Tanatside*, and HMS *Wensleydale* joined *Grenville* and *Ulster* on a Tunnel led by *Limbourne*'s Commander Byron Alers-Hankey. The opposing force, which had been providing distant cover to a coastal convoy, comprised the Kriegsmarine's Type 39s *T-22*, *T-23*, *T-25*, and *T-27* of the 4.Torpedobootflotille under Korvettenkapitän Franz Kohlauf, a respected leader with extensive night-fighting experience. A lighter horizon to seaward enabled Kohlauf to sight Commander Alers-Hankey's ships early, and he fired torpedoes at long range; all missed and escaped notice by the British. Soon after, *Tanatside*'s Type 271 detected the enemy at ten thousand yards. Alers-Hankey ordered star shells fired and when they burst overhead, Kohlauf withdrew at top speed. The slower Hunts fell behind, and a running gun battle developed between the four German torpedo boats and the two Fleet destroyers. *Grenville* and *Ulster* launched uncoordinated torpedo attacks that missed; their gunfire sprayed inaccurately. By contrast, *Grenville*'s Lieutenant Commander Roger Hill described "a large number of small splashes continuously round the ship, and occasional water on the bridge [from near misses]."[21] Enemy shells soon hit, igniting a large fire that forced *Grenville* to abandon the chase. *Ulster* continued the pursuit alone until hits also forced her to break off.

One might have thought that the first Tunnel to meet the enemy would spark timely analysis but that did not occur. Leatham only submitted his report to the Admiralty on 26 October, and analysts there did not circulate their commentary until 9 November. Meanwhile, the missions continued unabated, with Tunnels run on consecutive nights between 13–14 and 17–18 October. The objective was the German blockade-runner *Münsterland*, which special intelligence revealed was being sent up-Channel from Brest. She was an important target. Germany suffered a shortage of troop transports and had recalled vessels of suitable capacity from northwest Europe; of those, *Münsterland* was the most valuable.[22] The mid-October Tunnels featured the same mix of Hunts and Fleets, with HMS *Rocket* replacing *Ulster*, but they came up empty since *Münsterland* had yet to make her run from Brest.

This set the stage for the Tunnel of 22–23 October and the loss of the light cruiser HMS *Charybdis* and the Hunt-class destroyer *Limbourne*. The mission has been thoroughly dissected, but the haste with which it was mounted has been overlooked. Neither *Charybdis* nor her commanding officer, Captain George Voelcker, had participated in a Tunnel, nor, it appears, in an offensive sweep of any kind. Nonetheless, they were thrown into the mission with scant notice. *Charybdis* had returned to Plymouth from a Biscay patrol on the afternoon of 18 October; two days later she received notice for similar duty. Those plans were abruptly scrapped when special intelligence revealed that *Münsterland* would be escorted up-Channel on the night of 22–23 October. On the evening of 21 October, Leatham notified Voelcker that he would be leading a Tunnel the next night comprising *Charybdis*, the Fleet destroyers *Grenville* and *Rocket*, *Limbourne*, and three other Hunts. Thus, with just twenty-four hours' notice, Voelcker was thrust into command of a mission he had never undertaken, involving ships he did not know. Exacerbating an already poor situation, his second in command, Commander Walter Phipps replaced Alers-Hankey in *Limbourne* just two days prior to the operation and he had no Tunnel experience, either. Captain Voelcker's plan, presumably conceived with Leatham, was for *Charybdis* and the Fleet destroyers to tackle *Münsterland*'s escort, leaving the transport to the Hunts. Presumably because of his unfamiliarity with the others, Voelcker elected to steam in line-ahead even though it was considered unwieldy for more than five ships.

That night Kohlauf led five torpedo boats that made up a distant screen seaward of *Münsterland*. Again, he received plenty of warning of the enemy's approach, this time from shore-based radar; again, he had the benefit of a rising moon to seaward; and, again, he maneuvered his force into an advantageous tactical position. After sighting *Charybdis* against the brightened horizon, Kohlauf's force fired torpedoes at 0146, then wheeled into the loom of the coast.[23] Sweeping westward, Voelcker had ample notice the enemy was about. HMS *Talybont* and *Wensleydale*'s "Headache" operators—German linguists who monitored enemy radio transmissions—warned of chatter indicating that an enemy force was maneuvering in the vicinity, but Captain Voelcker did not grasp its significance. Then, at 0130, *Charybdis* detected a radar contact ahead at 14,000 yards. Voelcker did not relay that information, and because their Type 271 was masked ahead, none of the six destroyers picked it up themselves. "At 0130," Leatham later observed, "the situation in our force was that *Charybdis* knew that there was an enemy force 7 miles ahead and closing, but did not know its composition; while *Limbourne* and *Talybont* knew that there was a force of 5 ships (probably destroyers by the [R/T] procedure) in close vicinity, but did not know where."[24] Voelcker carried on until the range was down to four thousand yards, then fired star shell. At almost that exact instant a torpedo exploded against *Charybdis*' port side. As more torpedoes zipped through the British line *Charybdis* was hit again while another torpedo found *Limbourne*. The British force fell into disarray; *Grenville*'s Hill recalled "for the next fifteen minutes the ship was maneuvering to avoid collision." *Charybdis* went down quickly with 481 sailors, including Voelcker, while the fatally damaged *Limbourne* had to be dispatched by friendly torpedoes. A brief search by *Grenville* and *Rocket* failed to locate the enemy forces, which escaped without a shot being fired in their direction. *Münsterland* made it to Cherbourg, but was later sunk by artillery in the Dover narrows.

Kohlauf reflected that "fortune indeed smiled on us." That, and plenty of skill. Senior Kriegsmarine officers gushed, declaring that Fleet torpedo boats had proved invaluable: "With their speed, handiness, and small silhouette they appear to be suitable as sleuth hounds in our coastal routes in the Western Region, where they can strike with surprise and destructive force because of their strong torpedo armament." Their commanding officers "as well as their trusty and courageous ships' companies have achieved a further success by light surface forces that is unique in naval history." Predictably, analysis from Plymouth Command and the Admiralty was scathing. Lack of training, inexperience, and poor unit cohesion were identified as factors in the defeat, but the Admiralty staff fingered Voelcker. Rear Admiral Patrick Brind pronounced the final verdict: "Although it may be that the force was ill-assorted and ill-trained together, I do not consider the loss of the two ships can be attributed altogether to this cause. The loss was due in large part to a lapse of judgment by an experienced and very well thought of officer—the captain of *Charybdis*. To continue on a steady course toward unknown vessels almost head on was dangerous, particularly in view of the unfavorable conditions of light—a rising moon."[25]

REMEDIATION

The *Charybdis* debacle demonstrated that the situation in the western Channel must improve if the degree of sea control deemed necessary for the invasion was to be achieved. The prospect of enemy destroyers assaulting vulnerable troop transports unnerved Allied naval commanders. Leatham urged the establishment of a dedicated strike force at Plymouth comprising an "adequate division of fleet destroyers." The Admiralty agreed, replying, "In view of the importance attached to these operations, it is intended that in future you will have at your disposal, a more homogeneous and well-trained force."[26] Leatham wanted the force to comprise Tribal-class destroyers, with their powerful gun armament—six 4.7-inch and two 4-inch in twin turrets—but some preferred the J/K/N-class, with their greater torpedo strength: ten tubes vs. four. Leatham won out, likely because Tribals, the Royal Navy's answer to Japan's *Fubukis* described in Chapter 5, were more readily available. No matter what the choice, operational demands made any destroyer difficult to come by, but events in the last week of December 1943 transformed the situation. The destruction of the battle cruiser *Scharnhorst* in the Battle of North Cape on 26 December—a superlative example of how superiority at night could be decisive—enabled the Admiralty to allot a force "of ultimate strength" to Plymouth. Within 48 hours, they notified the Home Fleet, "It is intended to establish at Plymouth a force of two 6" and 2 5.25" cruisers and 8 heavily gunned Fleet destroyers with the intent of destroying enemy surface forces in the Channel and Bay areas, protection of convoys and craft engaged in amphibious exercises, and to intercept blockade-runners."

"As a first step," on 28 December they directed the Home Fleet to "sail 4 'Tribals' to Plymouth as soon as practicable."[27] That same day the situation improved further when the cruisers HMS *Glasgow* and HMS *Enterprise* sank the destroyers *Z-27*, *T-25*, and *T-26* in a running daylight battle in the Bay of Biscay. That enabled the Admiralty to reduce the Fleet destroyers earmarked for Plymouth from eight to five, precisely the number of Tribals in theater. In January HMS *Ashanti*, and her Canadian sisters *Haida*, HMCS *Athabaskan*, and HMCS *Iroquois*, arrived to form the Tenth Destroyer Flotilla (10 DF), which was really a half-flotilla led by a commander designated D10. HMCS *Huron* soon replaced *Iroquois*, while HMS *Tartar* joined as flotilla leader in February.[28] The two 6-inch cruisers never materialized, but the 5.25-inch light cruisers HMS *Bellona* and HMS *Black Prince* worked with the 10 DF on a rotating basis.

Although the 10 DF occasionally fulfilled some of the myriad tasks demanded of destroyers, it was first and foremost an offensive strike force and Leatham honed its readiness in that regard. That meant providing them with the best equipment, proper training, and experience. In terms of equipment, particularly all-important warning surface radar, they had mixed capability. *Ashanti*, *Athabaskan*, and *Tartar* had advanced Type 272 or 276 search radar mounted on stronger lattice foremasts and backed up by state-of-the-art Plan Position Indicator (PPI) displays. Upon arriving in Plymouth, *Haida*, *Huron*, and *Iroquois* had had the obsolescent Type 291 for search, but Leatham immediately had them fitted with Type 271Q with PPIs. This was a distinct improvement, but the aerial once again was located midships in the former searchlight position, with the nagging blind-spot ahead.

A key innovation was Action Information Organization (AIO). Over the winter of 1942–43 several destroyer captains complained of their difficulty processing the growing swell of tactical information from various sources. A senior operational commander remarked: "a stage has been reached where valuable information cannot be made use of by the Captain, at possibly a vital moment." The "haphazard installation" of instruments in destroyers aggravated the problem.[29] In response, in June 1943 the Admiralty convened an AIO committee. It conceived a "distillery" called the Action Information Center (AIC), which "would be complete in itself and will be fed with tactical and enemy information of all forms and from all sources. This centre will be vital to the efficient fighting—and defence—of the ship, and it will have to be under the direct control of a highly qualified officer."[30] Similar to the U.S. Navy's Combat Information Center, AICs were located on the signal deck beneath the bridge and included several warfare plots, the most important of which was a tactical plot fused with an ARL Table to present a continual, real-time tactical picture (see Chapter 3). An officer experienced in operations relayed this information to the bridge; if the captain sought a visual picture, a "view trunk" on the bridge provided a magnified window to the ARL tactical plot a deck below. A key component of the system was a dedicated Action Information Intercom (AID), which integrated tactical communications.

The optimal AIO setup was projected for future designs, but there was a pressing need to fit some version in existing destroyers. That process began in the autumn of 1943, but was hampered by the amount of work required, the lack of shipyard space, and the heavy demands on destroyers. That slowed installation in ships such as the Tribals; moreover, the ad hoc nature of the process meant that arrangements varied between ships. In the 10 DF, only *Tartar*—the flotilla leader—had robust AIO with a relatively complete AIC; the other destroyers received upgraded intercom systems, but their AIC arrangements varied. *Tartar's* AIC team was also the only one to get specialized instruction at a new AIO training center; the others had to hone their proficiency in sea training off Plymouth.[31] No matter how rudimentary the AIC setup or the extent of specialized training, however, AIO significantly improved the 10 DF's ability to unravel the confusion attending night action.

Other equipment enhanced capability. Navigation, notoriously difficult in the Channel, was simplified by the radio navigation aid QH-3, a variation of Bomber Command's "Gee" system that took cross bearings from shore-based transmitters. *Ashanti's* navigator recalled that it enabled him "to pinpoint their position virtually at the touch of a button."[32] Each ship also received Headache receivers to monitor enemy radio transmissions, TBS radio, the latest Identification Friend or Foe (IFF) gear, and updated Target Indication Units that improved synchronization between fire-control radar and fire-control systems. Finally, each destroyer received increased allocations of star shells, flashless cordite, and 4.7-inch tracer.[33]

Such upgrades were no guarantor of success—Voelcker's failure to appreciate the value of Headache is a prime example—and sea training proved critical. Soon after their arrival at Plymouth, Leatham sent *Haida*,

PHOTO 7.1. Hunters at speed: *Haida* maneuvering with *Ashanti* and *Bellona* or *Black Prince*. Relentless training ingrained the skill, cohesiveness, and confidence integral to success. Note *Haida*'s Type 271 lantern midships forward of the mainmast and the excellent arc of fire from the improved position of the pom pom. *DND R-1044*

Athabaskan, Iroquois, and *Ashanti* on a Tunnel operation with three Hunts. Predictably, problems arose and *Haida*'s DeWolf "strongly recommended that the Plymouth forces exercise night encounters." The Canadian knew what he was about; the RCN had emphasized night encounters in its prewar exercises, since the cover of darkness offered its small destroyer force the best odds against the commerce raiders that it expected to face.[34] Leatham, who had probably wanted to hammer home the necessity of training, laid on an extensive sea training program to prepare his ships for the demands of night operations, particularly the relatively new art of radar plotting. DeWolf's report from March 1944 reveals the scope of the training: "Night Encounter Exercises were carried out on the 3rd, 7th and 15th of March. The Force has also been exercised in manoeuvring, evading fighter attacks and radar-tracking. Destroyers have carried out high and low angle shoots, and towing evolutions. Comprehensive communications and plotting exercises were carried out on the 4th, 8th and 16th of March." The flotilla also fulfilled operational commitments, participating in three defensive patrols, two invasion exercises, a minelaying sortie, and an abortive sweep into the Bay of Biscay. The extent of this activity is confirmed by *Black Prince*'s Captain Dennis Lees, who recalled "days, weeks and months of dull, solid, slogging working up practices."[35]

Tactical publications were also available to coach destroyer officers on night fighting. Besides the doctrinal bibles *The Fighting Instructions* and *Destroyer Fighting Instructions*, handbooks such as the *Gunnery Review* and the *Guardbook for Fighting Experience* distilled lessons from previous actions. Interestingly, the night actions between the United States and Japanese navies in the Pacific received only marginal attention. Although Royal Navy liaison officers served in the Southwest Pacific, little information on those night surface actions seems to have been disseminated in the United Kingdom. Files on the actions filtered through the Admiralty and certain engagements were reviewed in *Fighting Experience*, but there does not appear to have been concerted effort to learn from the U.S. experience, and in the various analyses of the 10 DF's actions there is only one reference to fighting in the Pacific.

In the midst of their training the flotilla carried out several Tunnels "to see the elephant." No combat resulted, but the operations afforded valuable seasoning and a focus for training. On a sweep on 25 February 1944 confusion arose in mid-Channel when inexperienced radar operators mistook the approach of a friendly aircraft for an attack by Schnellboot. Then, when off the craggy Brittany coast, the flotilla opened fire on contacts classified as enemy destroyers, but which turned out to be small islands—they left haystacks burning in their wake. More serious problems arose on the night of 1–2 March when the flotilla leader, Commander St. John Tyrwhitt, in *Tartar*, twice turned his force away from what Headache indicated were torpedo attacks by enemy destroyers. The force returned to Plymouth after only a fleeting search for its adversaries; a determined hunt may have revealed five torpedo boats led by Kohlauf, who had stalked the Tribals and launched an unsuccessful torpedo attack. Leatham's subsequent analysis oozed disappointment: "depending on the circumstances, which only a commanding officer can gauge, evasive action may well have to be taken, but this does not mean that the main objective is to be missed as a result." The "main objective" still was to engage the enemy.[36] Tyrwhitt, who probably needed rest after two years of rigorous duty in the Mediterranean, was replaced as D10 after the failed sweep. The other captains took his replacement and Leatham's criticism to heart, embracing an aggressiveness that colored their subsequent operations. "My feeling at the time," DeWolf recalled, "was that we had failed to go after some ships, some echoes, so from then on in the back of my mind was if we ever get an echo, we're going after it. So, if we found one, we did; that was just natural fear of criticism."[37]

BAPTISM

After the actions in October 1943, Franz Kohlauf expressed concern that the enemy had been aware of his movements. Having encountered British forces on successive missions, Kohlauf complained: "The enemy doubtless knew about the operation in detail. Secrecy in Brest was very poor ... and establishments not involved knew about [their] commencement."[38] Kohlauf's hunch that the Allies had advance warning was correct, but the cause was special intelligence, not loose lips. Since 1941, British cryptographers had been able to decipher the Kriegsmarine's Home Waters Enigma, which covered surface ship movements. This "provided a reliable picture of the enemy's routines—showing which swept channels the convoys normally used, where they spent the night, the times at which they made and left harbor, and when and where they met their escorts."[39] It enabled analysts to divine the movements of escort forces such as Kohlauf's, giving the Allies a distinct advantage. It did not mean that interception was guaranteed, or that it would be successful; the October 1943 actions were proof of that. But it did mean that Leatham, who was cleared to receive Ultra information, could position his forces to best advantage.[40]

That very situation arose in the final week of April 1944. Signals intelligence revealed *Marine Gruppe West* planned to send a convoy westward from St. Malo to Brest, so Leatham laid on Tunnels. The first, on the night of 24–25 April, was unsuccessful. Force 26—the cruiser *Black Prince* with three Tribals—obtained radar contact with the convoy; however, operators watched their displays in dismay as the enemy, warned of Force 26's approach by coastal radar, found refuge in a harbor along the coast. Despite similar warning the next night, the Germans did not escape. At 0106 on 26 April, when Force 26—comprising *Black Prince, Haida, Athabaskan, Ashanti*, and *Huron*—was heading south eighteen miles off the French coast, the cruiser's F.V.1. radar monitor again indicated they were being held by German coastal radar. Kohlauf, providing distant cover to the convoy with three torpedo boats, had indeed received warning of Force 26, but anticipated that it would turn toward Brest to intercept the convoy, so he held his westward course. In fact, Force 26 never detected the convoy and had turned on its planned west-east sweep.[41] Coastal radar failed to alert Kohlauf of this alteration until 0201, by which time the opposing forces were on a collision course. *Black Prince*'s Type 272 tracked Kohlauf's ships as they steadily approached, until the enemy suddenly turned away. Rather than closing to fire torpedoes as he had done against *Charybdis*,

Kohlauf reversed course and cranked on full speed, presumably to draw Force 26 away from the convoy. Senior officers criticized his decision, noting that "a headlong attack launched without changing the westward course might have brought a hard fought for success. In situations that appear hopeless, the boldest decision leads in most cases to success."[42]

Kohlauf's "scoot" initiated a chase that tested new tactics adopted by 10 DF. Instead of line-ahead, Force 26 spread out in an open formation, with *Black Prince* in the center and a sub-division of two Tribals deployed at three thousand yards, 40 degrees off each bow of the cruiser. Based on the lessons of the *Charybdis* action, the formation reduced the cruiser's exposure to torpedoes while enabling her to maintain contact and illuminate the enemy. It also unleashed the Tribals' four forward 4.7-inch guns against the enemy and increased their freedom of maneuver. At 0220 *Black Prince* began illuminating at 13,600 yards; the Tribals opened up seven minutes later, firing as many as ten salvos a minute at the enemy. Although the range was long, with visibility impaired by German smoke, Type 285 provided accurate ranges, and since the torpedo boats did not use flashless cordite their gunflashes provided a convenient aiming point. *T-27*, the rearmost destroyer, was soon straddled, and at 0236 sustained hits that cut her speed to twelve knots and forced her to disengage toward the coast. *Haida* tracked the move, but DeWolf continued to pursue the two destroyers fleeing east. *T-27* fired torpedoes from long range as Force 26 swept by; all missed, but they caused *Black Prince* to turn away. For the rest of the night the cruiser shadowed the action from seaward "in case one of the enemy doubled back or one of our own destroyers required assistance."[43] *T-27* escaped into Morlaix.

With Lees on the periphery, command devolved on DeWolf, and with Leatham's earlier criticism in mind, he was determined to finish the job. Finding the range on Kohlauf's *T-29*, *Haida* and *Athabaskan* scored devastating hits at 0320, damaging the rudder, severing steam lines, and cutting her speed. An explosion on the bridge mortally wounded Kohlauf. By 0335, *T-29* lay still, engulfed in flames. In the meantime, *Ashanti* and *Huron* had pursued *T-24* to the east, scoring hits on her superstructure. Her speed was unaffected, and when she eventually escaped, the Tribals joined their mates. Spitting fire, the four Tribals circled the hapless *T-29*. When gunfire failed to sink her they turned to torpedoes, but incredibly all missed due to aiming errors or faulty drill—*Haida* fired from too close and her four torpedoes ran under the target.

Finally, at 0421 *T-29* slipped beneath the waves. As the Tribals re-formed, *Ashanti* collided with *Huron*, causing light damage to both ships.[44]

Neither side had opportunity to absorb lessons from the action before the two met again. Repairs to *Ashanti* and *Huron* kept them out of action until mid-May, and with *Tartar* in refit the operational load fell on *Haida* and *Athabaskan*, which had been scheduled for a rest. The *4.Torpedobootflotille* also suffered lingering effects. Unable to repair their battle damage in the small harbors they used for escape, on the night of 28–29 April, *T-27* and *T-29* stole down the coast for Brest. Special intelligence warned Plymouth Command of their movement, but nature gave a them a boost. Super-refraction caused by abnormal climatic conditions enabled shore-based radar at Plymouth to track *T-27* and *T-29* as they proceeded. At 0307, Vice Admiral Leatham ordered *Haida* and *Athabaskan*, in mid-Channel covering an invasion exercise, to speed south. Like air controllers vectoring fighters, Plymouth guided the Tribals to a perfect interception, and at 0359 *Athabaskan* gained radar contact at fourteen miles.

The Canadians were heading 170 degrees, almost due south, while the enemy was steering 260 degrees, almost due west. "[T]he ideal position for torpedo attack was developing automatically," a subsequent Admiralty analysis concluded. Commander DeWolf, however, rejected that option. "My first object was to prevent the enemy getting past to westward," he explained in his action report. With the 26 April action in mind, he wanted to avoid another chase. There were also small harbors into which the torpedo boats could duck, and they were only twenty miles from Ushant and the defense network at Brest. To mount an effective torpedo attack DeWolf would have to close to about five thousand yards, increasing the risk of detection by German coastal radar if not by the torpedo boats themselves, and opening the door for their escape. Weighing this on *Haida*'s bridge, DeWolf decided to close the enemy and engage with gunfire—"Straight on at 'em, I presume, Sir?" Pugsley had prompted Mountbatten in November 1940.[45]

Coastal radar gave *T-27* and *T-24* no warning, and at 0412 they were surprised when star shells burst overhead. Veering for the coast both fired torpedoes, but only three from *T-24* were launched to the correct side. When Commander DeWolf observed their turn away he "altered toward to avoid torpedoes, but limited the turn to thirty degrees to keep A arcs open." This enabled the 4-inch guns in X turret to illuminate the enemy, while the 4.7-inch guns of A and B mounts fired freely. But DeWolf probably also wanted to account for Type 271's blind-spot ahead; with Type 285 directing gunnery and Type 291 ineffective, he would have wanted Type 271Q's PPI available to plot both enemy ships. Before the turn was complete, a torpedo exploded against *Athabaskan's* hull, causing a large fire to erupt aft. *Haida* laid a smokescreen to cover her sister and then pursued the fleeing torpedo boats. Minutes later a massive explosion marked *Athabaskan's* demise.[46]

Dogging *T-27*, *Haida's* accurate shooting quickly settled the German vessel's fate. Two shells exploded on the water line while others burst against the superstructure. Losing speed, down by the bow, and with flames igniting ready use ammunition, her captain put *T-27* aground. DeWolf briefly sought *T-24*, which had disappeared eastward, before returning to where *Athabaskan* had last been seen afloat. In a courageous act, with daylight looming, DeWolf stopped in the shadow of the enemy coast for eighteen minutes rescuing forty-two sailors before reluctantly heading for Plymouth. Spitfires provided air cover from mid-Channel. German ships later picked up another eighty-five

PHOTO 7.2. The morning after. Upon *Haida's* return to Plymouth on the morning of 29 April 1944, a somber Admiral Ralph Leatham and Commander Harry DeWolf review the action that saw the loss of *Athabaskan* and the destruction of *T-27*. *DND PA180438*

Canadians, while six more made it home in *Haida*'s cutter. One hundred and twenty-eight died. *Athabaskan* was the largest warship loss suffered by the Royal Canadian Navy during the war.

SCRUTINY

In four nights the 10 DF had sunk two enemy destroyers against the loss of one of their own. Plymouth Command was relieved to achieve tangible success, but staff officers at the Admiralty's Directorate of Tactical and Staff Duties (DTSD), who had a professional responsibility to be coldly objective, were unimpressed. Reviewing the 25–26 April action, they criticized the escape of two enemy destroyers and deplored the failure to torpedo the motionless *T-29*. They were more scathing in regard to the second engagement, calling it "An unsatisfactory action in which, quite unnecessarily, we swapped a Tribal for an Elbing." Criticism of DeWolf's tactics was especially pointed. Captain St. John Cronyn, a senior warfare analyst, complained that the force "was so wrapped up in the picture of a gun action that it seems never to have contemplated the use of the torpedo." "It is deplorable to realize," he lamented, "that it was the German who sized up the situation."[47]

Cronyn had fired the opening salvo of a debate that endured through the summer of 1944. He and other analysts at the Admiralty believed that torpedoes were the most effective weapon in surface action, and tactical maneuvering should have the goal of using them effectively. In September 1944, after scrutinizing the 10 DF's entire record, the Admiralty issued a formal complaint to Leatham: "Considering our superiority in all but one of these [eight] actions, the enemy losses are disappointing. The main reason appears to be a general lack of appreciation of the use of the torpedo. It would appear that on several suitable occasions torpedo fire was withheld for no apparent reason, and on some occasions when torpedoes were fired results were disappointing."[48]

The Admiralty failed to appreciate the degree to which radar had changed the nature of night actions in littoral waters. Due to ship- and shore-based radar, it had to be assumed that the enemy had warning of the approach of any force. Moreover, the German "shoot-and-scoot" tactic, coupled with the high speed of their destroyers, made them difficult to engage with torpedoes let alone get in the first licks so important to success at night. This was an instance where technological development had outpaced the formulation of tactical doctrine, leaving sailors at the sharp end to work out tactics on the job. Innovation was required, and Commander Basil Jones, who succeeded Tyrwhitt as D10 on 15 March 1944, had both the knowledge to divine a solution and the will to see to its implementation. A gunnery specialist and seasoned destroyer officer, Jones had benefited from recent night-fighting experience in the Mediterranean. On the night of 15–16 April 1943 he led the British formation against two Italian torpedo boats, as related in Chapter 4. Notably, at the outset of that action he aggressively engaged the enemy bow-on with radar-assisted gunfire, using torpedoes as a weapon of opportunity.[49] After arriving in Plymouth in March 1944, Jones did not lead the 10 DF in action until June; nonetheless, he was instrumental in the implementation of new tactics. In this he was no doubt counseled by Leatham and Commander Reginald Morice, who ran Plymouth's destroyer staff, but it appears that the 10 DF's individual COs had little input; DeWolf, the flotilla's second in command, recalled no involvement.[50] Rather than an instance of overcentralization, this was more a case of "Here you go—get on with it," reflecting the command's confidence in its subordinates' training, skill, and initiative.

Jones explained his tactical thinking in a riposte to the Admiralty submitted after his flotilla wrapped up its service in European waters. "The problem presented to the 10th DF," he argued, "was, in the main, the destruction of a faster enemy who turned and fled while firing a greater number of torpedoes than we carried ourselves. The enemy's approved policy of turning to run gave him a position or opportunity of torpedo advantage." Radar-assisted torpedo-firing had proven unreliable so far, so Jones banked on gunnery:

> The four-gun forward armament of the 10th DF gave us a gun advantage over the enemy when advancing against his retreat. The fact that the enemy ships were faster than us made any delay in turning toward him generally unacceptable. The positively Elizabethan method of projecting torpedoes at right angles to own ship's fore and aft line is a cause of delay. Our policy was not to turn away nor intend to waste any time at all in closing the enemy.

Thus, aggressive use of radar-assisted gunnery presented the best opportunity to engage first effectively. Moreover, it satisfied what one historian has described as "the ancient master law of tactics—being superior at the point selected to deliver the decisive blow."[51] Summing up his philosophy, Jones emphasized, "This attitude was not due to lack of appreciation of the use of the torpedo, which is undeniably a more effective weapon than the gun if it can be applied."[52]

To gain maximum gunnery advantage, the 10 DF had "to press on into the enemy during his turn away." Although line-ahead made for easier station keeping, Jones thought it unsuitable for the head-on encounters prevalent in the Channel, since ships were prevented from entering action together and those at the head of the line obstructed the radar of those behind. Instead:

> It was desirable that all destroyers should have their forecastle guns bearing, their Radar unimpeded ahead, and ships capable of individual action to comb enemy torpedoes. Only a reasonably broad and shaken-out line of bearing formation could satisfy these conditions. It was realized that cruising at night for lengthy periods in such a formation was a strain as regards station keeping, although the P.P.I. removed much of the strain. Accordingly Line Ahead for comfort, and Line of Bearing for action, was the order of the day.[53]

Although Jones did not participate in either engagement, the open formation utilized in the 25–26 April action and DeWolf's preference for guns three nights later demonstrated that his philosophy had taken hold.

AFFIRMATION

Athabaskan's loss reduced the 10 DF to four Tribals, but reinforcements arrived throughout May. The British destroyers HMS *Eskimo* and *Javelin* were of the Tribal- and J-class respectively. Both had undergone significant modernization, including upgraded surface radar and some elements of AIO. The two other newcomers were stalwarts of the Polish navy. The Polish destroyer *Piorun* was a former Royal Navy N-class, almost identical to *Javelin*, while *Blyskawica* was unique. Built to special design by a British shipyard in the mid-1930s, she was larger than her flotilla mates and was reputed to be the fastest Allied destroyer in the theater. Both Polish destroyers had Type 271 mounted midships, but neither appears to have been fitted with AIO. Satisfying the gunnery criteria that shaped the flotilla, all four newcomers had either twin 4.7- or 4-inch turrets forward. Their commanding officers were all seasoned destroyer captains, but as we have seen, that did not always translate into success, particularly when innovative tactics were being employed. Their late arrival—*Blyskawica* joined just two weeks before the invasion—left little time for group training or an opportunity to foster mutual trust and confidence. To compensate, Leatham and Jones concentrated their four battle-hardened veterans—*Tartar, Ashanti, Haida,* and *Huron*—in the 19th Division with the newcomers in the 20th under *Blyskawica*'s Komandor Konrad Namiesniowski.

The challenges associated with absorbing new ships into the 10 DF were trivial compared to those confronting the enemy. Kriegsmarine destroyer strength had shrunk through steady losses, and on the eve of the invasion just three destroyers and one torpedo boat of the 8.*Zerstörerflotille* were at Bordeaux to attack Neptune's western flank. The 4.Torpedobootflotille had been dissolved after the losses inflicted by 10 DF, while three torpedo boats of 5.Torpedobootflotille operated out of Le Havre to the east.[54] The 8.Zerstörerflotille was a mixed bag. Z-32 and Z-24 were modern Type 36A destroyers, popularly known as "Narviks." Displacing 3,000 tons, capable of thirty-eight knots, and armed with five 5.9-inch guns and eight 21.7-inch torpedo tubes, Narviks were larger, faster and packed a heavier punch than British destroyers.[55] The smaller, slower *ZH-1*—the captured Dutch destroyer *Gerard Callenburgh*—was equivalent to an interwar Royal Navy "A-I" class vessel. The fourth ship was the 10 DF's old adversary, *T-24*; smaller and slower than her flotilla mates. All four possessed superb torpedo-control gear, which made it their most potent weapon. Radar remained of 1940 vintage with limited search capability, but the Narviks had modifications to overcome the 30-degree blind-spot astern that afflicted *ZH-1* and *T-24*. Machinery breakdowns were also common.[56] The range of capabilities meant that if the vessels were to operate as a cohesive unit they would have to conform to *T-24*'s performance, which could deprive the Narviks of their speed advantage.

Training was another deficiency. The persistent transfer of officers and ratings to the U-boat arm caused disruption and increased the requirement for sea training, but that was increasingly difficult to achieve since fuel was scarce and Allied airpower threatened whenever ships put to sea. This curtailed group training to the point where, before the June action, the flotilla's ships never went sea together. Readiness languished and senior leadership could do little to compensate. Since April 1944 the flotilla had been led by Kapitän zur See Theodor von Bechtolsheim, a respected destroyer officer who had commanded destroyer *Karl Galster* in numerous night actions, including against Mountbatten's 5 DF in November 1940. He had since served as chief of staff to the *Führer der Zerstörer*, so he was abreast of the latest trends in destroyer warfare. All captains save *ZH-1*'s had commanded their ships in action, and *T-24*'s Kapitänleutnant Wilhelm Meentzen had a unique edge in having survived two battles against the 10 DF.

Among myriad concerns, the Germans' radar disadvantage loomed large. Kriegsmarine officers suspected that the Allies had blind-fire capability, and confirmation came from *Athabaskan*'s survivors. *T-24* had rescued about fifty *Athabaskans*, but Meentzen observed that "They are disciplined and reserved. They make very few statements. All that we can learn is that they come from a Canadian destroyer of the 'Tribal class' that had been sunk by a torpedo hit." More ruthless interrogators, uninhibited by respect for fellow men of the sea, pried out valuable intelligence. "These facts," the staff reported, "clearly confirm the enemy's substantial superiority in the field of radar and in particular in blind fire." With palpable bitterness, the author complained, "It is most unfortunate that British gunnery, which learned how to shoot from the old Imperial Navy and copied its methods several decades ago, has now achieved such a significant lead because of technical superiority."[57]

Apprehension would have been more deeply rooted had the 8.Zerstörerflotille realized that the enemy knew their intentions on an almost real-time basis. Raids by Allied air forces and French partisans had disrupted German communications, forcing them to often use radio instead of secure landlines. On D-Day afternoon when *Z-32*, *Z-24*, and *ZH-1* were ordered from the Gironde to join *T-24* at Brest, their route and timetable were decrypted and forwarded to the Plymouth Command within hours.[58] Leatham initially intended the 10 DF intercept them en route to Brest, but when he learned of a major U-boat offensive against the invasion from Biscay bases he elected to deploy a Coastal Command antishipping strike force. This apparent "dilly-dallying" irritated the First Sea Lord, Admiral of the Fleet Andrew Cunningham, scrutinizing operations from the Admiralty war room. "C-in-C Plymouth appears to be afflicted with infirmity of purpose," Cunningham groused in his diary. "A most mistaken sense of values for which we may pay dearly. I will give him snuff in the morning although it's little to do with me."[59] Cunningham's unease reflects the seriousness with which Operation Neptune planners viewed the Kriegsmarine destroyer threat; nonetheless, it is hard to fault Leatham's logic. His destroyers would be vulnerable to U-boats in the Bay, and they also could be mistakenly engaged by the Allied aircraft blanketing the area; better that airmen knew that every contact could be considered hostile. Moreover, the Coastal Command force at Leatham's disposal had proven its antishipping capability. The strike wing, flying Bristol Beaufighters armed with 25-pound armor-piercing rocket projectiles and 20-mm cannon—all part of the British No. 144 and Canadian No. 404 squadrons—had ravaged German coastal shipping off Norway before moving south to bolster the western flank of the invasion.[60]

At 2030 on D-Day, thirty-one wave-hopping Beaufighters intercepted Kapitän zur See Von Bechtolsheim's destroyers off St. Nazaire. The fighter-bombers attacked out of the low sun, and wheeling aircraft, exploding flak, and rocket trails crowded the horizon. In the face of fierce antiaircraft fire, the Beaufighters inflicted only light damage, and one was shot down. A follow-up strike the next morning shot up the three destroyers as they approached Brest. In the end the attacks were not as decisive as Leatham hoped, and after a thirty-six-hour delay for minor repairs Von Bechtolsheim headed out to attack the invasion. The 10 DF lay in wait. Allied intelligence analysts kept a close eye on Brest, and Enigma decrypts on 8 June revealed the 8.Zerstörerflotille was deploying to attack invasion shipping between Portsmouth and Normandy, putting into the shelter of Cherbourg at dawn. Beyond the zerstörerflotille's intent, special intelligence divulged its precise course and speed, enabling Leatham to position Force 26—the 10 DF—across their path.[61]

At 0114, 9 June the 19th Division was zigzagging on a base course of 255 degrees at twenty knots, while the 20th Division bore 000 degrees at two miles, slightly abaft the 19th's beam. Both steamed in staggered line-ahead with radar unimpeded. The sky was overcast with intermittent rain, a light breeze blew out of the southwest, and the sea was calm. Visibility was variable at one to three miles with a rising moon. *Tartar's* Type 276 detected a contact directly ahead, bearing 241 degrees at ten miles. At 0120, after the AIC developed an accurate plot, Jones increased to twenty-seven knots, and two minutes later ordered Force 26 into loose line abreast to increase freedom of maneuver and make it easier to avoid torpedoes. The 19th Division's commanding officers maneuvered their destroyers with practiced ease, but not the 20th Division. Although *Blyskawica's* signal log recorded Jones' order, Namiesniowski later wrote, "I have no recollection of having had this signal reported to me." Here, again, arose the friction that chronically besets night fighting. Instead of eight destroyers confronting the enemy in line-of-bearing, only the 19th Division was in the intended formation, while the 20th maintained line-ahead behind *Blyskawica*.

MAP 7.2. 9 JUNE 1944 ACTION

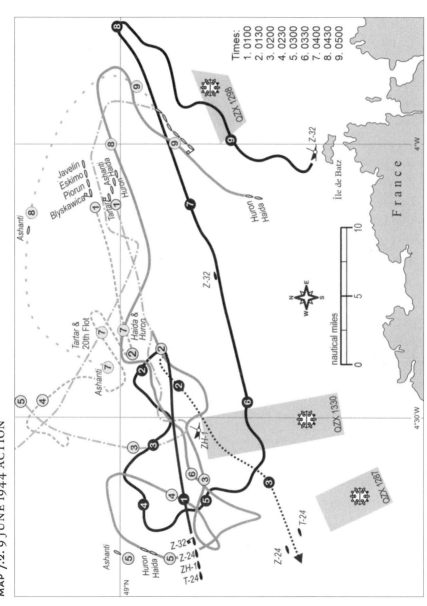

Times:
1. 0100
2. 0130
3. 0200
4. 0230
5. 0300
6. 0330
7. 0400
8. 0430
9. 0500

QZX 1298

QZX 1330

QZX 1287

Z-32

France

Île de Batz

4°W

4°30'W

nautical miles

0   5   10

N
W   E
S

Javelin
Eskimo
Piorun
Błyskawica

Ashanti
Haida

Tartar

Huron

Ashanti

Tartar &
20th Flot

Haida &
Huron

Ashanti

ZH-1

Z-32

Huron
Haida

Huron
Haida

Z-32
Z-24
ZH-1
T-24

Ashanti

Z-24

T-24

49°N

Von Bechtolsheim expected to encounter Allied warships. Shortly before midnight he forecast dim prospects. "I cannot imagine that the British are going to allow us to complete our passage to Cherbourg unscathed," he recorded in his war diary. "On the other hand, we can make life unpleasant for them in the Bay of the Seine. I expect to encounter enemy destroyers no later than reaching the Channel Islands." It did not take that long. After Z-32 gained a fleeting radar contact, Commander Jones' maneuvering exposed the 19th Division to moonlight. Von Bechtolsheim reported "Enemy in sight" and wheeled south to place his ships against the darker horizon. Unhappily, his navigator warned that a minefield lay down that track, so he hauled around to seaward, ordering his ships to fire torpedoes on the turn. Z-32, Z-24 and ZH-1 each unleashed four "eels" (torpedoes) at the charging 19th Division. The 20th Division was still undetected to the north, but Headache and the flexibility afforded by line abreast enabled Jones' division to avoid them.[62]

Jones sought a close-range "pell-mell battle," which is precisely what broke out after the 19th Division opened fire, followed closely by the 20th. The German destroyers were turning to port, crossing the bows of the 19th Division at about thirty-five hundred yards in the order Z-32, ZH-1, Z-24, T-24. *Tartar* initially engaged Z-32, hitting her four times, but when she sped off northward Jones left her to the 20th Division and joined *Ashanti* against ZH-1, then Z-24. *Haida* fired on Z-24 then joined *Huron* against T-24.[63]

Had the 20th Division been spread out in line-of-bearing, Z-32 would have been in dire straits. As it was, the usual problems of line-ahead arose. The two sides exchanged fire and Z-32 sustained hits forward; however, before the 20th division could press its advantage, *Blyskawica*'s Headache operator reported the enemy's "*Toni Dora*," the code term for "fire torpedoes." Rather than turning toward the enemy, Komandor Namiesniowski hauled around to starboard. The rest of the division followed, its captains thinking they were carrying out a torpedo attack—*Eskimo* and *Javelin* each snapped off four that missed. Instead of pursuing Z-32, Namiesniowski continued away from the action. He offered no explanation, although he may have been concerned about mistakenly engaging Jones' division; Leatham thought AIO would have clarified the situation. Seventeen minutes passed before the division returned to the sound of the guns, and its ships had no further impact on the action.[64] If nothing else, the incident underlines the value of seasoning. It had been long-standing Royal Navy night-fighting doctrine to turn toward enemy torpedoes to maintain contact. Experience would have demonstrated the necessity of this maneuver, and training would have made it easier to accomplish. However, since the 20th Division was new to the flotilla, there was neither the time nor the opportunity. Leatham and Jones surely knew this, leaving one to wonder what might have transpired had the more-experienced DeWolf, the 10 DF's second-in-command, led the 20th Division.

Having escaped the clutches of the 20th Division, Von Bechtolsheim found himself immersed in a close-quarters brawl more resembling the Age of Sail than modern naval combat. Heading west in hope of re-forming the force, Z-32 sighted *Tartar* close on her port quarter at 0138 and scored three hits on the Tribal's bridge superstructure, killing several of the crew and slightly wounding Jones. Communications were disrupted, and Jones temporarily lost control of Force 26.[65] Z-32 also suffered in the exchange, causing Von Bechtolsheim to disengage. As he withdrew, *Ashanti* brought Z-32 under fire; however, before she could inflict decisive damage, smoke from *Tartar*'s fires cloaked the enemy. As *Ashanti* sorted out the confusion, ZH-1 emerged from the smoke. She had followed Z-32 around to port at the outset to the action, but lost contact with Z-24 and T-24; therefore, when she sighted ships, her commanding officer chose not to engage in case they might be friendly. Under no such illusion, *Ashanti* and *Tartar* struck first. Among a deluge of hits, 4.7-inch rounds penetrated ZH-1's machinery spaces bringing her to a dead stop. When the veil of smoke and steam lifted, ZH-1 lay unmasked before her assailants. Despite her own damage, *Tartar* raked the enemy from point-blank range while *Ashanti* put a torpedo into her, blowing off most of the bow. At 0240, ZH-1 blew up in a massive blast that reverberated across the western Channel.[66] Thirty-nine died.

Z-24 and T-24, the remaining German destroyers, were initially targeted by *Haida* and *Huron*. *Haida* first engaged Z-24 from about four thousand yards. DeWolf recalled that the target, "just then turning away, very quickly started to make smoke and zig zag at fine inclinations. Some ten or fifteen salvoes were fired at this target, and several hits were possibly scored before another target was observed to the left."[67] Z-24 suffered damage to her bridge, machinery spaces, and forward gun mount. Having observed Z-24 turning away to westward, T-24 followed, assuming they were following Z-32.[68] The battle now took a familiar turn for *Haida* and *Huron*. On 26 April they had become embroiled in a long chase; now they pursued Z-24 and T-24 southwest.

Conditions were unfavorable, and DeWolf later reported that the enemy destroyers "were engaged with the wind dead ahead and rain squalls were frequent. Cloud base was never more than 1000 feet and often as low as 500 feet. Consequently illumination was poor and star shell were generally half burned before they effected any illumination whatsoever. The enemy made excellent use of smoke throughout and continuously took avoiding action thus making spotting at times well nigh impossible."

Despite the challenges, the Tribals, racing southwest at thirty-two knots, may have overhauled the enemy had fortune not intervened. At 0150 the plots in both Canadian warships indicated that the Germans had entered the Allied minefield QZX-1330, which they were under orders to avoid. While *Haida* and *Huron* skirted the field, *Z-24* and *T-24* steamed through unaware and unscathed. By the time the Tribals resumed direct pursuit they had fallen nine miles behind and soon lost radar contact. At 0214, his position "with regard to own forces and remainder of the enemy obscure," DeWolf abandoned the chase to seek out *Tartar*.[69] *Z-24* and *T-24* continued for Brest.

Typical of most night actions, the situation had become confused, with both commanders unsure of the position of their own forces, let alone the enemy's. To regain control of Force 26, at 0237, Jones directed Force 26 to concentrate on *Tartar*. To the west, Von Bechtolsheim made similar efforts and headed "on a southern course in order not to get too far away from the battle area."[70] Meanwhile, *Haida* and *Huron* proceeded cautiously northeast, unaware of *Tartar*'s exact location. Darkness magnified uncertainty. At 0223 both Tribals obtained a firm radar contact bearing 032 degrees at six miles. Since their plots indicated that *Tartar* should bear 040 degrees, they thought it might be her. IFF provided no confirmation, but they assumed that *Tartar*'s set might be damaged. DeWolf wrote,

> Made identification by light and ordered Plot to carry out radar search for other ships which might be concentrating. Ship in sight replied to our signal by light, but his signals were unintelligible. Main armament was brought to the ready and the challenge made, but the reply was again unintelligible. I still considered it might be *Tartar* with damaged signaling gear and [wounded] personnel. The ship made smoke and turned away to the west and south but was not plotted by Radar and range was opened to 9000 yards before this move was appreciated.[71]

Z-32 was equally in the dark. Seeking Z-24 and T-24, Von Bechtolsheim reported: "Individual shadows are sighted. Exchanges of recognition signals by blinker gun, and even by night identification signal, do not lead to any identification. The fact that, despite German recognition signal interrogation, these shadows do not fire, however, causes me to make the decision not to use my weapons."[72] Z-32 hurried away to the northwest, gradually working around to an easterly course. *Haida* and *Huron* followed, and at 0254 any doubts about the contact's identity were removed when star shells revealed the distinctive silhouette of a Narvik. Both Tribals opened fire.

The two commanders had different objectives. DeWolf's was the destruction of the enemy; Von Bechtolsheim's was to survive as a threat to the invasion. Von Bechtolsheim wrote,

> As I am operating alone with Z-32 I can achieve nothing against the overwhelmingly superior enemy, particularly as I must first load my spare torpedoes and there is no longer sufficient gun ammunition for a prolonged engagement. Proceeding eastwards must remain the aim! As I can no longer reach Cherbourg before first light, I decide to make for St. Malo. I hope to get the other destroyers to re-join during the passage and before reaching the Channel Islands. I therefore order the following: "Break through to the East. Goal St. Malo, join me."[73]

DeWolf had a better grasp of the situation. In response to his query about the earlier explosion that had lit up the horizon, *Ashanti* informed him that it was an enemy destroyer blowing up. "Nice work," DeWolf replied. He had chased two other destroyers toward Brest and knew from experience that they were unlikely to rejoin the battle. That left the destroyer bathed in star shells, which DeWolf knew he could focus on exclusively.

The Tribals overcame poor shooting conditions as they pursued Z-32. Von Bechtolsheim mistakenly assumed the accurate fire was due to flares that had been dropped from aircraft. They scored several hits, but before those had any effect minefield QZX-1330 again intervened. Z-32 unwittingly entered the minefield at 0311. *Haida* and *Huron* again had to avoid it and by the time they resumed the chase Z-32 was ten miles to the southeast. Minutes later, they lost radar contact—one can imagine their frustration at twice being thwarted by a friendly minefield. DeWolf, under instructions to re-form on *Tartar*, could have broken off the pursuit without fear of criticism; in fact, a senior officer present on *Haida's* bridge recommended that he do just that. Rear Admiral Harold Reid, the third senior-most Canadian officer who had fortuitously picked that night to ride in *Haida*, urged DeWolf to abandon the chase. It is testament to DeWolf's confidence and determination that he rejected Reid's advice. Although he thought the enemy might escape into Morlaix, he continued the chase with *Huron* faithfully keeping pace, displaying the initiative that can tilt the balance in night action.[74]

Von Bechtolsheim was equally resolute. Although Z-32 had suffered damage it was "not serious enough to compel avoiding a fresh engagement." He predicted that this would occur near the Channel Islands, and believed Z-24 and T-24, which he thought trailed only twelve miles astern, would link up by then. That prospect was crushed when at 0420 they reported they were twenty-five miles to westward heading for Brest. "With a heavy heart," Von Bechtolsheim lamented, "I must therefore decide to abandon the mission ordered. I cannot achieve a breakthrough to the East with Z-32 alone in these circumstances. Whether I will be able to break through to the West remains to be seen. Moreover, I suspect the presence of warships, as there is a shadower to the Northwest."[75] If Z-32 had been equipped with more effective search radar, Von Bechtolsheim would not have had to speculate what lay to the northwest;

PHOTO 7.3. Z-32 hard on the rocks of Ile de Batz after her running fight with the 10 DF. Her main armament remains fixed to the starboard quarter as stoic testament of her persistence against her assailers. *DND-CN-6870*

nonetheless, his suspicions proved correct. Having regained contact, *Haida*'s and *Huron*'s Type 271Q indicated they were overtaking Z-32, and twenty minutes later the range began to drop rapidly. It soon became apparent that the enemy had turned back westward, so the Canadians turned south to cut him off. Meanwhile, Commander Jones concentrated the rest of Force 26 six miles to the north to cover any escape eastward. Z-32 was trapped. At 0444 *Haida* and *Huron* opened fire from seven thousand yards. Von Bechtolsheim, believing that he was under attack from two cruisers—German sailors commonly mistook the large Tribals for cruisers—he turned toward the coast, returned fire, and launched his remaining torpedoes. They missed, and although several 5.9-inch shells burst close to the Tribals, they inflicted no damage. The Canadians were unsure of the effect of their gunnery, but Von Bechtolsheim attested: "The ship is constantly caught by hits. The way things are going, my running won't last long." *Blyskawica* briefly joined in from the north, but it was unclear whether her shooting had caused any damage. At 0500, Z-32's port engine broke down and hits put her forward turret out of action. Von Bechtolsheim attempted to escape under the loom of the coast, but more hits knocked out the starboard engine. Realizing the end was at hand, he ran his destroyer aground on the rocky shore of Ile de Batz.[76] After almost drowning, Von Bechtolsheim made it ashore.

The defeat dashed any German hopes of using their most powerful ships to intercede against the western flank of the invasion. Not only had they lost Z-32 and ZH-1, but the damage to Z-24 took weeks to repair and there was little T-24 could do alone. The Germans attributed the defeat to their poor state of training, the withdrawal of Z-24 and T-24, and overwhelming odds. Von Bechtolsheim won praise as a "daring, experienced and resolute commander" who "brought honor to the destroyer arm."[77] The Allies celebrated. Leatham described the action "as one of the very few which has been fought by large and fast destroyers at night during this war." Its "successful outcome was due primarily, in my view, to the correct and immediate action of Commander Jones . . . to force close action, while at the same time avoiding the enemy's torpedo fire." The 19th Division's other commanding officers also received praise, as did many of their sailors. Yet, there still was some unease. Admiral Bertram Ramsay, the naval commander for Operation Neptune, grumbled to his diary, "I wanted all to be sank." Leatham and the Admiralty staff thought that but for the 20th Division's "inexcusable" turn away, the 10 DF would have routed the enemy.[78] That seems dubious. All that may have changed is that Z-32's fate may have been settled sooner, preventing the mauling of *Tartar*.

Having achieved its primary mission to win control of the western Channel, the 10 DF applied its hard-won night expertise against German shipping over the remainder of the invasion summer, running close inshore in the Chops and the Bay of Biscay. New challenges emerged, but, again, sailors absorbed lessons and in two actions in August the 10 DF annihilated lightly defended convoys. The highlight was *Iroquois'* masterful use of her newly fitted AIO, which featured her captain, Commander James Hibbard, conning the force into action from his AIC—likely the first time that occurred in the European theater, and a portent of the future. Even so, the success garnered criticism: "I want my Captains on the bridge," one admiral insisted.[79]

LEGACY

The night fighting in the Chops of the Channel fostered professional introspection. Since the actions occurred in the dawn of the radar age they became the focus of study in the postwar Royal Navy, and, by extension, the Royal Canadian Navy. The *Charybdis* fiasco and April 1944 actions received detailed analysis in a confidential study by the Admiralty's DTSD, which repeated its previous criticisms of the 10 DF's use of torpedoes, but failed to mention Leatham's and Jones' robust rebuttals. The Royal Navy tactical school also dissected the actions, apparently with some criticism of Jones' tactics. In 1950 Jones justified his decisions in the Royal Navy's professional journal, *The Naval Review*, relating them to Horatio Nelson's desire "to shock and to surprise." He also took solace in late-war Fleet destroyer designs, such as the Battle, Weapon, and *Daring* classes, which resurrected the Tribals' powerful forward gun armament—a step that Jones thought would promote the "end-on" approach that he had championed.[80] Even with DTSD's partiality and Jones' sensitivity, the scrutiny reflects a high degree of naval professionalism. The night fighting in the Chops of the Channel introduced new challenges—challenges that would grow more complex as navies ventured further into the radar age. It was, after all, still-developing technology with much yet to learn. As this story of night fighting over five decades of conflict demonstrates, achieving the desired capability to fight effectively in the dark required determined study of lessons learned, coupled with the search for innovative solutions. The 10 DF's experience in doing just that signaled a way ahead.

CONCLUSION

In late 1943, then–Capt. Arleigh Burke was mentoring a young ensign in tactical decisionmaking. Burke, a seasoned veteran of Cape St. George and Empress Augusta Bay, asked the young man to tell him "the difference between a good officer and a poor one." The ensign hesitated, and then described the importance of "aggressiveness, technical proficiency, command presence, knowledge of human nature" and other leadership qualities that had been emphasized in his education. Burke listened patiently and then said, "The difference between a good officer and a poor one is about ten seconds."[1]

Prior to the late nineteenth century, night was regarded as a time for sneak attacks or accidental encounters. The invention of torpedoes and searchlights transformed this perception and created an apparent opportunity for weaker forces to use asymmetric means, supplemented by stealth and surprise, to overcome stronger foes. The Imperial Japanese Navy and Germany's Kaiserliche Marine, in particular, pioneered night-combat tools and techniques, in the belief that such engagements would provide them an advantage. In practice, however, night combat proved difficult to manage. Through World War I most night actions occurred by accident rather than design, and results failed to meet expectations. As this book has endeavored to show, there were many reasons for this disconnect between anticipated potential and reality. Certain patterns emerge from the forty-year-long process by which navies gained competency, if not mastery of the night, and from these patterns in turn come lessons that are highly relevant for navies of today.

TEN SECONDS

The night actions described in this book illustrate the truth of Burke's insight on several levels. In situations where both sides were alert, prepared for battle, and ready for action, timing was a crucial factor. Seconds separated victory from defeat at Cape St. George, in Cdr. Frederick Moosbrugger's triumph at

Vella Gulf, and during Commander Basil Jones' achievements in the Chops of the Channel and the Mediterranean. When one side was surprised or caught unprepared—as at Port Arthur, Cape Matapan, and Savo Island—the other had a nearly insurmountable advantage. But Burke's aphorism was about more than just catching the enemy unaware; it also emphasized the importance of fast decisions—especially those informed by training, doctrine, and an effective plan. Look at the Beta Convoy to see the price of hesitation.

The speed at which one side could make sense of a situation and exploit it was crucial. Drill and practice familiarized crews with the challenges of working together and coordinating their operations in the face of nocturnal confusion and sudden surprises. Effective tactics increased the lethality of ships and weapons. Successful doctrine improved the ability of formations to fight as a cohesive unit, rather than as individual ships. The speed with which a commander could act on a night contact was largely determined by their level of preparation and investments made long before the decisive moment. Realistic training often provided a decisive advantage.

The Japanese experience shows the value of such training. Their emphasis on *nikuhaku-hitchū* ("press closely, strike home") led Japanese flotilla craft to make determined attacks during the Russo-Japanese War. Tactical refinements and technological developments in the interwar period of 1919–39 enabled the Imperial Japanese Navy to sharpen its night combat potential further. Skilled lookouts, well-drilled crews, and superior weapons ensured that throughout much of World War II the Japanese could continue to successfully employ *nikuhaku-hitchū*, seizing momentary opportunities to devastate enemy formations with well-aimed torpedo salvos, as they did at the battles of Java Sea, Tassafaronga, Kula Gulf, and Kolombangara. Thorough preparation gave Japanese officers the ability to anticipate, react, and deliver attacks more quickly than many of their adversaries.

The British navy made similar investments preparing for night combat in the interwar period and stands out for developing night-battle tactics at every level, even for the largest ships. The contrast with Italy's Regia Marina was stark. Believing that darkness was best-suited for harassing, attritional attacks by small craft, the Regia Marina's major units were ill-prepared for nocturnal action at the outset of World War II. It took the loss of a heavy cruiser division at Cape Matapan to convince Supermarina of the urgent need to prepare large ships for night action.

The way navies conceptualized night action affected their success. Each navy developed tactics, doctrine, and techniques for night combat that reflected its national perspective and strategic circumstances. All navies prepared for *the* decisive clash of battle lines because they believed this was essential to achieving sea control; but not all prepared for a sustained campaign—a war to dominate the vital maritime lines of communication that enabled land and air forces to be supplied. The convoy battles in the Mediterranean, the fighting in the Solomons, and the Tunnel Operations in the English Channel were efforts to control lines of communication, to use the sea to secure victory on land; night-combat capabilities were essential to such a campaign. The navies that were better able to adopt this perspective, to subordinate victory in a specific battle to success in a broader campaign, won World War II.

The shift in Japan's perspective over time is illustrative in this regard. Vice Admiral Tōgō Heilhachirō attacked Port Arthur in February 1904 to ensure the successful landing of Japanese troops on the Korean peninsula. He used his naval power to secure a broader strategic goal—control of the sea lanes between Japan and the continent. By 1942, Japan's emphasis on decisive battle had altered its strategic orientation and encouraged the belief that victory at sea was an end unto itself. The navy was "preoccupied with the decisive battle engagement" and "indifferent to the protection of vital sea communications."[2] Thus, after the success of Vice Admiral Mikawa Gunichi at the Battle of Savo Island, he preserved his ships, but failed to destroy the U.S. transports that were vital to the survival of the Marines on Guadalcanal. During the ensuing campaign, naval power played a crucial role, but it was logistics that decided the issue. Tōgō thought in terms of a campaign; in contrast, Mikawa emphasized battle. Their approach to night combat reflected these differences.

The U.S. Navy, ostensibly the logical heir to Mahan and his emphasis on decisive battle, instead recognized that in the Pacific it needed to fight a sustained campaign, and that required secure lines of communication. When presented with the opportunity to seek a decisive night action with the Japanese during the 1944 Marianas Campaign, Adm. Raymond A. Spruance and Vice Adm. Willis A. Lee—perhaps because of his razor-thin margin of victory at the Second Naval Battle of Guadalcanal—declined. They prioritized preserving the amphibious forces over destroying the Japanese fleet.[3] Many in the U.S. Navy criticized this choice, but their commander in chief, Adm. Ernest J. King, told Spruance "that he had done exactly the correct thing."[4]

Success in night combat required making the most of Burke's fleeting ten seconds. It required a means to process information effectively so that commanders and crews could penetrate the darkness, overcome the confusion, and, as Capt. Wayne P. Hughes, so eloquently framed it, "attack effectively first."[5] Initially, that meant thorough preparation and indoctrination so that distributed groups of torpedo craft could employ surprise and confusion to make a decisive attack. By "striking boldly" and "pressing home," early torpedo craft could overcome the defenses of larger ships, launch their weapons from close range, and secure crippling hits. Later, as the tactics of night combat became increasingly sophisticated, it involved obtaining, as the British dubbed it, the "instantaneous production of maximum output."[6] Linear formations—"following Father"—appeared to offer the best means of achieving this because they helped retain cohesion while simultaneously concentrating firepower. In battle, such formations had numerous successes, such as Kapitän zur See Michelsen's raid in the Dover Strait and Captain William Agnew's destruction of the Beta Convoy. However, tactical forms alone were insufficient. Rigorous training and effective doctrines were required to harness the potential of modern ships and technologies in night combat and concentrate decisive force at the outset of an action. The interwar British and Japanese navies excelled in this regard, but even their tactics and training were insufficient to overcome the confusion that ensued after firing began.

Ultimately, mastering the night required effective training and indoctrination coupled with new systems that maximized the potential of human cognition so that officers and sailors could quickly make better sense of the world around them by overcoming the limitations of their eyes and ears. Radar was the most important sensor in this regard, but VHF/UHF radio, direction-finding, and tactical signals interception also played key roles. To be most effective, the information streams from those sensors had to be synthesized into a continuously updated "picture" that could serve as the basis for action. That task was too great for any one human mind. The cognitive burden had to be distributed to a team, and such teams were the heart of the U.S. Navy's CIC and Royal Navy's AIO. When coupled with centimetric radars and visually intuitive PPI displays, CICs and AIOs could ameliorate many of the traditional challenges of night combat. The "pictures" they provided led to revolutionary tactics that allowed distributed formations to concentrate their firepower at night as never before. That ability is what led to definitive success at Surigao Strait and in the Chops of the Channel.

Other navies also recognized the potential of such sensors. However, Japan's reliance on optics and pyrotechnics led it down a different evolutionary path, and by the time the Japanese discovered that electronic means offered better results, it was too late for them to keep pace with the Allies. The Germans leveraged shore-based radars to great effect, giving their forces in the English Channel a capability that exceeded their shipboard sensors; but they never created an integrated information system on board their vessels comparable to the Allied systems, and they suffered in combat accordingly.

TECHNOLOGY AND HUMAN POTENTIAL

The impact of the CIC and AIO illustrates the symbiotic nature of technology and tactics. Technology is most effective when it enables new levels of human potential. Technology is not an end unto itself, but a means of harnessing new, more sophisticated forms of organization, coordination, and combat power. The core technologies of this narrative—torpedoes, torpedo boats, destroyers, searchlights, radio, and radar—were important because they transformed naval combat by making it possible to fight effectively at night. As the previous chapters have recounted, the world's major navies endeavored to adopt these technologies, integrate them into their force structures, and hone their skills to fight in the dark. None of this was easy. Navies regularly misjudged the potential of these technologies, and even if they grasped how best to use them, implementation posed a slew of new challenges. The record demonstrates that some approaches were more effective than others, but it also reinforces the subtext of Burke's counsel: it is not the ten seconds themselves that matter, but what an officer does with those seconds before they tick away. Of all the factors in combat, human decisionmaking is the most important. It is people that "dominate battle."[7]

In this regard, a crucial factor in the equation for success was the ability to learn. This may seem a given, but history proves that it is not. People and organizations require feedback. To be successful, a feedback system needs two things. The first is a process to gather information about what happened. That may seem to come naturally, but the major investments in intelligence services show how much work is required to do it well. The second is intellectual honesty. Navies needed to be honest in accepting the mistakes they make and honest in evaluating the results they achieve. The U.S. and British navies had effective systems to gather combat intelligence and disseminate battle lessons; they used those lessons to improve the way they integrated technology and tactics to increase their effectiveness at night.[8] The way the Italian navy was able to improve its night-combat skills in the face of persistent defeat was due in large part to a cold-blooded pragmatism. In contrast, the Japanese navy was less willing to take systemic approaches and learn at scale from combat experience. According to his memoir, when Captain Hara Tameichi returned to Truk from the bloody First Naval Battle of Guadalcanal he received praise and recognition, but the Combined Fleet staff was "not interested in learning from [my] recent battle experiences."[9]

Officers who could be confident in their ships, their subordinates, and their view of a developing action could be bold, like Commander Harry DeWolf in the Channel on 9 June 1944; Rear Admiral Tanaka Raizō at the Battle of Tassafaronga; Captain Ernesto Pellegrini, who skippered the Italian cruiser *Scipione Africano* on her run to Taranto in July 1943; and Capt. Frank Walker at the Battle of Vella Lavella. They could risk much and exploit uncertainty. The confusing nature of night combat rewarded this behavior, and those forces which—through training, tactics, doctrine, or technological advantage—could recognize fleeting opportunities, capitalize on them, and seize the initiative usually emerged victorious. As American naval historian and strategist Alfred Thayer Mahan so aptly put it, "[G]ood men with poor ships, are better than poor men with good ships" and "this lesson, . . . [in] our own age, with its rage for the last new thing in material improvement, has largely dropped out of memory."[10] New technologies made naval night combat possible, but, as the preceding chapters have repeatedly demonstrated, it took well-trained officers and sailors to win in the dark. That was true in the twentieth century; it is no less true today.

APPENDIX

Increasing effectiveness at night, along with other factors such as more capable aircraft, better sensors, and superior intelligence, led to more night battles being sought. During World War I, 18 of 47 (38 percent) night surface engagements were intentional.[1] As the nature of combat at sea changed, the number of intentional encounters in World War II grew to more than 83 percent (80 of 97). Improved intelligence was an important factor in this shift. Just 40 percent of night actions in World War I stemmed from information that one side obtained about the other's intentions or movements. In World War II, that number grew to almost 70 percent. The increase in deliberate efforts to seek night battle reflected the importance of controlling lines of communication. In World War II, 43 percent of night actions involved an attack against a harbor, beachhead, or convoy. Many of these actions were described in chapters 4, 5, and 6; they include the attack on the Beta Convoy, the night actions off Guadalcanal, and the Battle of Surigao Strait. In World War I, just 12 percent of night battles were fought over a harbor, beachhead, or convoy; far more—44 percent—involved general patrols, raids, or interceptions.

TABLE A.1. WORLD WAR I SURFACE ENGAGEMENTS

CATEGORY	COUNT	% TOTAL	DAY	% DAY	NIGHT	% NIGHT
Intentional	80	60.2%	62	72.1%	18	38.3%
Intelligence	66	49.6%	47	54.7%	19	40.4%
Nonconsensual	99	74.4%	57	66.3%	42	89.4%
Surprise	47	35.3%	34	39.5%	28	59.6%
Total	133		86		47	
Harbor Attack	9	6.8%	8	9.3%	1	2.1%
Beachhead Attack	3	2.3%	3	3.5%	0	0.0%
Convoy Attack	5	3.8%	3	3.5%	2	4.3%
General Attack	59	44.4%	45	52.3%	14	29.8%
Encounter	57	42.9%	27	31.4%	30	63.8%
Total	133		86		47	

TABLE A.2. WORLD WAR II SURFACE ENGAGEMENTS

CATEGORY	COUNT	% TOTAL	DAY	% DAY	NIGHT	% NIGHT
Intentional	135	82.8%	55	83.3%	80	82.5%
Intelligence	109	66.9%	43	65.2%	66	68.0%
Nonconsensual	132	81.0%	53	80.3%	79	81.4%
Surprise	72	44.2%	18	27.3%	54	55.7%
Total	163		66		97	
Harbor Attack	14	8.6%	10	15.2%	4	4.1%
Beachhead Attack	11	6.7%	3	4.5%	8	8.2%
Convoy Attack	45	27.6%	15	22.7%	30	30.9%
General Attack	54	33.1%	23	34.8%	31	32.0%
Encounter	39	23.9%	15	22.7%	24	24.7%
Total	163		66		97	

COMPARING NIGHT AND DAY

These numbers show the increasing amount of information available to naval officers, and the growing potential for navies to use their forces to achieve strategic ends. In World War II, operational commanders proved more capable at positioning their ships effectively to intercept the enemy, destroy convoys, and protect amphibious forces. However, no matter how precise that positioning, the basic truths of night combat remained. It was always quick, confusing, and fraught with uncertainty and risk. Even the best-fought night actions later in the war, such as those off the coast of Brittany in June 1944 or around the Philippines later that year, demonstrate this fact.

NOTES

PREFACE

1. Ronald McKie, *The Survivors* (Indianapolis, IN: Bobbs-Merrill Company, 1953), 71. He is describing the 28 February–1 March 1942 Battle of Sunda Strait.

2. Meir Finkel, *On Flexibility: Recovery from Technological and Doctrinal Surprise on the Battlefield* (Redwood City, CA: Stanford University Press, 2011), 24.

INTRODUCTION

1. Carl von Clausewitz, *On War,* indexed, trans. and ed. Michael Howard and Peter Paret (Princeton: Princeton University Press, 1984), 119.

2. The 25 July 1943 "Battle of the Pips," for example, apparently involved a U.S. Navy task force firing at "dense flocks of . . . dusky shearwaters"—seabirds that sleep in large groups on the surface of the sea and often create radar returns when they take flight. See Fern Chandonnet, ed., *Alaska at War: 1941–1945, The Forgotten War Remembered* (Fairbanks, AK: University of Alaska Press, 2008), 31–32.

3. "Fighting Instructions, March 1653," BCW Project: British Civil Wars, Commonwealth & Protectorate, 1638–1660, http://bcw-project.org/texts/fighting-instructions.

4. Julian Stafford Corbett, *Fighting Instructions, 1530–1816: Publications of the Navy Records Society, Vol. XXIX*, Project Gutenberg, 2005, https://www.gutenberg.org/ebooks/16695.

5. See also Julian S. Corbett, ed., *Signals and Instructions 1776–1794 with Addenda to Vol. XXIX* (London: Navy Records Society, 1908).

6. Corbett, *Signals and Instructions,* 306, 310.

7. Mario Navi-Mocenigo, *Storia della Marina veneziana da Lepanto alla caduta della repubblica* (Rome: Ministero della Marina, 1935), 343–44.

8. This was the pre-dreadnought *Royal Sovereign* laid down in 1891 and already an antique by 1902, not the 15-inch gunned ship laid down in 1914 that served in both World Wars.

9. Quoted in Bryan Ranft, ed., *Technical Change and British Naval Policy* (New York: Holmes & Meier, 1977), 25.

10. Katherine C. Epstein, *Torpedo: Inventing the Military-Industrial Complex in the United States and Great Britain* (Cambridge, MA: Harvard University Press, 2014), 60–65.

11. Norman Friedman, *British Destroyers: From the Earliest Days to the Second World War* (Barnsley: Seaforth, 2009), 28.

12. Arthur Hezlet, *Electronics and Sea Power* (New York: Stein and Day, 1976), 11.

13. Hezlet, 10–12.

14. J. B. Murdock, "The Naval Use of the Dynamo Machine and Electric Light," U.S. Naval Institute *Proceedings* (August 1882 8/3/21, 343–85), 343.

15. J. Heinz, "Electric Searchlight at Sea," U.S. Naval Institute *Proceedings* (October 1894 20/4/72, 763–82), 766.

16. Heinz, 780.

17. Commencement address to graduating ensigns of the Midshipmen's School USS *Prairie State*, 7 June 1941. David Sears, *Duel in the Deep: The Hunters, the Hunted and a High Seas Fight to the Finish* (publication pending).

18. Quoted in Jon T. Sumida, *In Defence of Naval Supremacy: Finance, Technology, and British Naval Policy, 1889–1914* (New York: Routledge, 1993), 14.

19. Theodore Ropp, ed. by Stephen S. Roberts, *The Development of a Modern Navy: French Naval Policy, 1871–1904* (Annapolis: Naval Institute Press, 1987), 177.

20. Ropp, 162–63.

21. Tami Davis Biddle, *Rhetoric and Reality in Air Warfare: The Evolution of British and American Ideas about Strategic Bombing, 1914–1945* (Princeton: Princeton University Press, 2002).

22. See Jon Tetsuro Sumida, *Inventing Grand Strategy and Teaching Command: The Classic Works of Alfred Thayer Mahan Reconsidered* (Baltimore: Johns Hopkins, 1997), 61.

23. Quoted in Kevin D. McCranie, *Mahan, Corbett, and the Foundations of Naval Strategic Thought* (Annapolis: Naval Institute Press, 2021), 108.

24. Quoted in McCranie, 112.

25. This was particularly true in the Imperial German Navy and Imperial Japanese Navy. See Dirk Bönker, *Militarism in a Global Age: Naval Ambitions in Germany and the United States before World War I* (Ithaca, NY: Cornell University Press, 2012), 104, and also Sadao Asada, *From Mahan to Pearl Harbor: The Imperial Japanese Navy and the United States* (Annapolis: Naval Institute Press, 2006), 182.

26. Quoted in McCranie, 114.

CHAPTER 1. STUMBLING IN THE DARK

1. Russian Naval General Staff, *Russko-Iaponskaia voina 1904–1905 gg.* [The Russo-Japanese War 1904–1905] (7 vols.; St. Petersburg: various publishers, 1912–1917), 1:218–220 [hereinafter RNGS].

2. All dates are given according to the western (Gregorian) calendar; at this time Russia still used the Julian calendar, which was thirteen days behind the Gregorian.

3. Julian S. Corbett, *Maritime Operations in the Russo–Japanese War, 1904–1905*, 2 vols. (Annapolis: Naval Institute Press, 1994), 1:324.

4. Newton A. McCully, *The McCully Report: The Russo–Japanese War, 1904–05*, ed. Richard A. von Doenhoff (Annapolis: Naval Institute Press, 1977), 160–61.

5. *The Russo-Japanese War, 1904–1905. Reports from Naval Attachés* (Nashville, TN: The Battery Press, 2003), 320 [hereinafter *Attaché Reports*]. Robert Gardiner ed., *Conway's All the World's Fighting Ships 1860–1905* (New York: Mayflower Books, 1979), 205–7.

6. *Organizatsiia artilleriiskoi sluzhby na sudakh 2-i eskadry Flota Tikhogo okeana* [Organization of the Gunnery Service aboard the Vessels of the Second Squadron of the Fleet of the Pacific Ocean] (St. Petersburg: Tipografiia Morskogo Ministerstva, 1905), 15. On night sights in general, see H. Garbett, *Naval Gunnery* (originally published 1897; reprint edition: East Ardsley: S.R. Publishers Ltd., 1971), 207–9.

7. D. K. Brown, "The Russo-Japanese War: Technical Lessons as Perceived by the Royal Navy," in *Warship 1996* (London: Conway Maritime Press, 1996), 66–77. On page 74, Brown states that Japanese torpedoes were not fitted with gyroscopes, but RNGS, 1:225,

notes that unexploded Japanese torpedoes found after the attack on Port Arthur did have them, as did Russian torpedoes, according to Iu. L. Korshunov and G. V. Uspenskii, *Torpedy Rossiiskogo flota* [Torpedoes of the Russian Navy] (St. Petersburg: Gangut, 1993), 28. However, one recent article claims that "far from all" torpedoes were so fitted at the time of Tsushima (Aleksandr Aleksandrov, "Srazhenie v Tsusimskom prolive" [Battle in the Tsushima Strait], *Floto-Master*, no. 4, 2005, 6–16).

8. Eric Lacroix and Linton Wells II, *Japanese Cruisers of the Pacific War* (Annapolis: Naval Institute Press, 1997), 778.

9. RNGS, 7:189.

10. Arthur Hezlet, *Electronics and Sea Power* (New York: Stein and Day, 1975), 21, 53, 55.

11. Jean Rey, *The Range of Electric Searchlight Projectors*, translated by J. H. Johnson (New York: D. van Nostrand Company, 1917), 59–61.

12. Based on a review of numerous recent Russian publications on individual classes of warships.

13. Hans Lengerer and Lars Ahlberg, *Capital Ships of the Imperial Japanese Navy 1868–1945*, vol. I: *Armourclad Fusō to Kongō Class Battle Cruisers* (Zagreb: Despot Infinitus d.o.o., 2019), 117, 122, 126, 186.

14. L.G. Beskrovnyi, *Armiia i flot Rossii v nachale XX v. Ocherka voenno-ekonmicheskogo potentsiala* [The Army and Navy of Russia at the Beginning of the Twentieth Century: A Study of Military-Economic Potential] (Moscow: Nauka, 1986), 206; Evans and Peattie, 554n9.

15. RNGS, 1:163.

16. Evans and Peattie, 70, 550n44.

17. Corbett, 1:474, quoting Tōgō's "Battle Instructions," article I.d.

18. Evans and Peattie, 67–74, 519.

19. Evans and Peattie, 46, 49.

20. Evans and Peattie, 91–2, 553–54n105; see S. O. Makarov, *Discussion of Questions in Naval Tactics*, trans. by John B. Bernadou (Annapolis: Naval Institute Press, 1990), 165–73.

21. *Attaché Reports*, 210, report dated 28 September 1904.

22. Brown, 74, without citing a source.

23. Corbett, 1:412.

24. *Morskoi Ustav 1899* [Naval Regulations 1899] (St. Petersburg: Naval Ministry, 1899), 43–44. See also Nicholas Papastratigakis, *Russian Imperialism and Naval Power: Military Strategy and the Build-up to the Russo-Japanese War* (London: I. B. Tauris & Co. Ltd., 2011), 52.

25. V. V. Iarovoi, "Parokhod 'Velikii Kniaz' Konstantin" [The Steamer *Velikii Kniaz Konstantin*], *Gangut*, no. 27 (2001), 31–38.

26. Bernd Langensiepen and Ahmet Güleryüz, *The Ottoman Steam Navy 1828–1923* (London: Conway Maritime Press, 1995), 6, 162.

27. A. M. Petrov et al., *Oruzhie Rossiiskogo flota (1696–1996)* [Weapons of the Russian Fleet (1696–1996)] (St. Petersburg: Sudostroenie, 1996), 68.

28. A. A. Remmert, "Kak pol'zovat'sia boevymi fonariami dlia otrazheniia minnykh atak" [How to Use Battle Lamps for Repelling Torpedo Attacks], *Morskoi Sbornik*, vol. CCCV, no. 7 (July 1901), unofficial section, 113–28, at 113.

29. J. N. Westwood, *Witnesses of Tsushima* (Tokyo: Sophia University; Tallahassee, FL: The Diplomatic Press, 1970), 103; Tōgō's "Battle Instructions," articles I.3.d and IV.9, quoted in Corbett, 1:475, 488.

30. McCully, 161.

31. For a summary of the issues besetting the Russian navy's personnel, see Papastratigakis, 53–54.

32. RNGS, 1:153–54.

33. McCully, 77, 161, 247.

34. Based on a review of Suliga, 3–4, which lists the names of commanders and their periods aboard individual boats.

35. McCully, 160–61.

36. This analysis is based on a list officers commanding destroyers and torpedo boats at the war's start and at Tsushima, compiled from Japanese Naval General Staff, *Opisanie voennykh diestvii na more v 37–38 gg. Meidzi (v 1904–1905 gg.)* [A Description of Military Operations at Sea in the Years 37–38 of the Meiji Era (in 1904–1905)], translated by the Russian Naval General Staff, 4 vols. (St. Petersburg: Naval Ministry, 1909), 1:21–5, 4:15–21 (hereinafter JNGS). This list was then checked against the tabular career summaries of Japanese naval officers at the website http://admiral31.world.coocan.jp/e/p_index.htm; out of a total of 121 officers, 41 (35 percent) could be positively identified.

37. *Attaché Reports*, 18, report dated 19 February 1904.

38. *Attaché Reports*, 346, report dated 6 May 1905.

39. Papastratigakis, 247, 255, 264.

40. M. V. Bubnov, "Port-Artur. Vospominaniia o dieiatel'nosti Pervoi Tikhookeanskoi eskadry i morskikh komand na beregu vo vremia osady Port-Artura v 1904 g." [Port Arthur: Recollections of the Activities of the First Pacific Ocean Squadron and the Naval Command Ashore during the Siege of Port Arthur in 1904], *Morskoi Sbornik*, part 2: vol. CCCXXXIV, no. 5 (May 1906), unofficial section, 29–56, at 32; RNGS, 1:175.

41. Stark's report dated 8 February 1904, quoted in RNGS, 1:182.

42. RNGS, 1:191.

43. RNGS, 1:156–160.

44. Bubnov, 32.

45. Frederick McCormick, *The Tragedy of Russia in Pacific Asia*, 2 vols. (New York: The Outing Publishing Company, 1907), 1:65.

46. Corbett, 1:75.

47. Corbett, 1:89; Ian Nish, "Japanese Intelligence and the Approach of the Russo-Japanese War," in *The Missing Dimension: Governments and Intelligence Communities in the Twentieth Century*, ed. Christopher Andrew and David Dilks (Urbana: University of Illinois Press, 1984), 17–32, at 29.

48. JNGS, 1:23.

49. Corbett, 1:94.

50. The following account is based on Corbett, 1:94–100; JNGS, 1:55–57; and RNGS, 1:198–223. Russian time is used, which was 55 minutes behind Japanese time.

51. Corbett, 1:95.

52. This follows RNGS, 1:199; however, other sources claim that the Russian destroyers did see the Japanese, *e.g.*, [British] Committee of Imperial Defence, Historical Section, *Official History (Naval and Military) of the Russo-Japanese War*, 3 vols. (London: HMSO, 1910–1920), 1:58 [hereinafter *Official History*].

53. *Attaché Reports*, 15, report dated 14 February 1904; Evans and Peattie, 554n7.

54. Corbett, 1:96.

55. *Attaché Reports*, 16, report dated 14 February 1904.

56. The number of torpedoes fired is taken from Corbett, 1:98, who had access to a Japanese navy confidential history, but Evans and Peattie, 97, say twenty torpedoes, the *Official History*, 1:58, says eighteen torpedoes, while JNGS, 1:55–7, provides details of sixteen torpedo launches. Estimates of hits: *Attaché Reports*, 9, 16, reports dated 10 and 14 February 1904.

57. *Kasumi* has been credited with the torpedo hit on *Pallada* (Watts and Gordon, 238).

58. Corbett, 1:100n.

59. RNGS, 1:203n, 1:217.

60. Watts and Gordon, 238.

61. Corbett, 1:100n.

62. Evans and Peattie, 98.

63. RNGS, 1:225.

64. The following account is based on RNGS, 1:467–475, and JNGS, 1:97–102.

65. Bubnov, 101.

66. Matusevich's report, reproduced in Russian Naval General Staff, *Russko-Iaponskaia voina 1904–1905 gg. Diestviia floty. Dokumenty* [The Russo-Japanese War 1904–1905. Operations of the Fleet. Documents] (St. Petersburg: Naval Ministry, 1911), Part III, Book 1, Issue 2, 43 [hereinafter *Dokumenty*].

67. JNGS, 1:99.

68. RNGS, 1:472.

69. RNGS 1:471, 474 and JNGS 1:98–100. Corbett, 1:148, gives total Japanese casualties as twenty-four killed and wounded, while the *Official History*, 1:82, reports seven killed and eight wounded.

70. *Attaché Reports*, 48, report dated 12 March 1904; Brown, 69.

71. R. M. Mel'nikov, *"Riurik" byl pervym* [*Riurik* was the First] (Leningrad: Sudostroenie, 1989), 88, 90; A. B. Shirokorad, *Entsiklopediia otechestvennoi artillerii* [Encyclopedia of National Artillery] (Minsk: Kharvest, 2000), 356.

72. Bubnov, 101n.

73. Addendum to Matusevich's report, *Dokumenty*, III:1:2:46.

74. *Official History*, 1:83; Tōgō Kichitarō, *Naval Battles of the Russo-Japanese War* (Tokyo: Gogakukyokwai, 1907), 25.

75. *Official History*, 1:82.

76. Corbett, 1:279, 298–9.

77. The following account is based on RNGS, 2:166–172, 175–180, 193; JNGS, 1:178–182; and *Attaché Reports*, 107, report dated 4 July 1904.

78. RNGS, 2:271–5.

79. Torpedo caliber based on Lengerer and Ahlberg, 192.

80. See RNGS, 3:47–145; JNGS, 1:218–27; Corbett, 1:405–12.

81. Corbett, 1:412; V. Ia. Krestianinov and L. A. Kuznetsov, *Eskadrennye bonenostsy tipa "Poltava"* (St. Petersburg: Gangut, 2020), 232.

82. Corbett, 2:105–26.

83. Westwood's *Witnesses of Tsushima* provides an invaluable Russian perspective on the voyage to Tsushima and the battle; Corbett, 2:217–333, remains the best overall account of the battle in English.

84. For the intelligence received by Rozhestvenskii, see Constantine Pleshakov, *The Tsar's Last Armada: The Epic Voyage to the Battle of Tsushima* (New York: Basic Books, 2002), *passim*. The measures taken to protect his squadron are described in the diary of an unknown Russian officer serving aboard the battleship *Orël*, translated and reproduced in [British] Intelligence Department No. 807, August 1907, "Reports from Naval Attachés, &c.," vol. IV, Appendix VI. See also RNGS, 6:113.

85. *Orël* diary, 220, 222; P. A. Vyrubov, *Desiat' let iz zhizni russkogo moriaka, pogibshego v Tsusimskom boiu (v pis'makh k ottsu)* [Ten Years in the Life of a Russian Sailor Who Died in the Battle of Tsushima (in Letters to his Father)] (Kiev: Tipografiia 1-i Kievskoi artel pechatnogo dela, 1910), 131–32.

86. *Organizatsiia artilleriiskoi sluzhby*, 10, 15, 17, 18.

87. J. N. Westwood, *Russia Against Japan: A New Look at the Russo-Japanese War* (Albany: State University Press of New York, 1986), 144.

88. *S eskadroi admirala Rozhestvenskogo: Sbornik statei* [With the Squadron of Admiral Rozhestvenskii: A Collection of Articles] (originally published in Prague, 1930; reprinted: St. Petersburg: OBLIK, 1994), 88–89; Westwood, *Witnesses*, 232, 242; V. Iu. Gribovskii and I. I. Chernykov, *Bronenosets "Admiral Ushakov"* [Battleship Admiral Ushakov] (St. Petersburg: Sudostroenie, 1996), 183.

89. Gribovskii and Chernykov, 185; Pleshakov, 211. One Russian source notes that at Tsushima Nebogatov's ships "did not switch on their searchlights, *following the experience in this regard of the Port Arthur squadron*" (my emphasis); "Zakliuchenie Sliedstvennoi komisii po vyiasneniiu obstoiatel'stv Tsusimskago boia" [Conclusions of the Investigative Commission for Clarifying the Circumstances of the Tsushima Battle], *Morskoi Sbornik*, vol. CDI, no. 8 (August 1917), unofficial section, 1 -42, at 16. This report, originally prepared immediately after the war, was not made public until 1917.

90. This paragraph is based on Evans and Peattie, 111–14.

91. Eric Grove, *Big Fleet Actions: Tsushima–Jutland–Philippine Sea* (London: Arms and Armour Press, 1991), 26.

92. Evans and Peattie, 557n44. Although the Japanese considered linked mines highly secret and made no mention of them in published descriptions of the battle, the Russians seem to have had some knowledge of them; see "Story of the Operations of the Russian Fleet at Pt. Arthur as Told by one of the Commanders," *Scientific American*, vol. 92, no. 10 (11 March 1905), 206–08; the diagram at the bottom of page 206 provides an almost exact representation of how the linked mines were supposed to work. The "commander" in question was von Essen, who was interviewed by the magazine while returning home by way of the United States.

93. The following is based on RNGS, 7:178–200; "Zakliuchenie," 13–42; V. Iu. Krest'ianinov, *Tsusimskoe srazhenie 14–15 (27–28) maia 1905 g.* [The Tsushima Battle, 14–15 (27–28) May 1905] (St. Petersburg: Galeia Print, 1998), 122–6; and Corbett, 2:294–331.

94. RNGS, 7:181, 183.

95. Russian time, which was 18 minutes behind Japanese time (RNGS, 7:121n.1). All times should be considered approximate; the Russians were reconstructing events from memory, while the Japanese torpedo craft were probably unable to keep accurate records of times and targets.

96. RNGS, 7:190.

97. Quoted in V. V. Arbuzov, *Bronenosnyi kreiser "Admiral Nakhimov"* (St. Petersburg: Korabli i srazheniia, 2000), 109.

98. Quoted in Corbett, 2:302–3.

99. According to Corbett, 2:305, and Evans and Peattie, 122, *Sisoi Velikii* was torpedoed by the Fourth Destroyer Division after 0200 on 28 May, but Russian sources state that she was torpedoed before midnight.

100. "Zakliuchenie," 23.

101. Evans and Peattie, 122.

102. Corbett, 2:304.

103. Aleksandrov, 7–8.

104. "Zakliuchenie," 23.

105. Corbett, 2:309–10. The totals of torpedoes expended during the night has been calculated by subtracting those fired at the crippled battleship *Kniaz Suvorov* during the daylight action; see Corbett, 2:265, 2:270–1, 2:291. However, these numbers should be regarded as approximate, as it is impossible to establish accurate figures from the available accounts.

106. Arthur J. Marder, *From the Dreadnought to Scapa Flow*, vol. 1 (London: Oxford University Press, 1961), 329; Evans and Peattie, 128.

107. Evans and Peattie, 555n26.

108. *Attaché Reports*, 138, report dated 3 August 1904; McCully, 162, 247.

109. *Attaché Reports*, 210, report dated 28 September 1904.

110. Hezlet, 14.

111. Unfortunately there is as yet no full-length biography of von Essen or a detailed study of his role in rehabilitating the Baltic Fleet; for an overview, see N. B. Pavlovich, ed., *The Fleet in the First World War*, vol. I: *Operations of the Russian Fleet* (New Delhi: Amerind Publishing Co. Pvt. Ltd., 1979), 59–67. For an in-depth study of the first few years of von Essen's command, see E. F. Podsobliaev, "Shkola admirala Essena" [The School of Admiral Essen], *Gangut*, no. 12-*bis* [*i.e.*, no. 13] (1997), 117–27.

112. Pavlovich, 85–123.

113. Evans and Peattie, 103–31, 151.

CHAPTER 2. TACTICS OF FRUSTRATIONS

1. The National Archives [hereinafter TNA], ADM 137/4324, German Documents. *Entwurf zur Seekrieganleitung*, vol. 2, *Torpedobootstaktik* (Reichsdrucke: Berlin, 1914), 47 [hereinafter *Torpedobootstaktik*]. Emphasis in original. All translations are by the author.

2. Otto Groos, *Der Krieg in der Nordsee*, 5 vols. (Berlin: E. S. Mittler, 1923–25), 3:65, quoting Carl's account of the action. "154β" refers to a grid square on the German naval chart of the North Sea.

3. Groos, 3:65–66; Admiralty, Naval Staff, Training and Staff Duties Division, *Naval Staff Monographs (Historical)*, vol. 3, 178–80 [hereinafter *Monographs*]; James Goldrick, *Before Jutland: The Naval War in Northern European Waters, August 1914–February 1915* (Annapolis: Naval Institute Press, 2015), 201.

4. Erich Gröner, *German Warships, 1815–1945*, vol. 1, *Major Surface Vessels* (Annapolis: Naval Institute Press, 1990), rev. ed., xi; "Rundschau in allen Marinen Monatsrundschau: Deutschland" [hereinafter RaMMD]: *Marine Rundschau* 21, no. 1 (January 1910): 103–4.

5. "Stand und Ausslichen der Torpedowaffe im fremden Marinen," *Marine Rundschau* 19, no. 10 (October 1908): 1088; "Die Wirkung des neuen englischen 7000 Yard-Torpedos auf den Entwurf von Kreigschiffsneubauten," *Marine Rundschau* 20, no. 2 (February 1909): 192.

6. "Das Torpedoboot," *Marine Rundschau* 21, no. 8 (August 1910): 957.

7. *Torpedobootstaktik*, 73–84, 85–94.

8. For German planning before the war, see Ivo Lambi, *The Navy and German Power Politics, 1862–1914* (Boston: Allen & Unwin, 1984).

9. *Torpedobootstaktik*, 47.

10. "Das Torpedoboot," 947–64.

11. Gröner, 1:173, 175; Norman Friedman, *Naval Weapons of World War One: Guns, Torpedoes, Mines and ASW Weapons of All Nations, An Illustrated Directory* (Barnsley, UK: Seaforth, 2011), 336–37; Roger Branfill-Cook, *Torpedo: The Complete History of the World's Most Revolutionary Naval Weapon* (Barnsley, UK: Seaforth, 2014), 240; ADM 137/4327, German Documents. *Taktische Befehle der Hochseeflotte 1914*, 31.

12. *Torpedobootstaktik*, 77.

13. *Taktische Befehle*, 34–36, 42–86.

14. Gröner, 1:173, 175.

15. Gröner, 1:178.

16. ADM 189/39, *Annual Report of Torpedo School, 1919*, 208.

17. ADM 137/2085, *Harwich Force, Destroyers*, 1.

18. ADM 137/2085, *Destroyers*, 1; U.S. Navy Office of Naval Intelligence (ONI), *Monthly Intelligence Bulletin*, 15 June 1921, 157.

19. John Brooks, *The Battle of Jutland* (Cambridge, UK: Cambridge University Press, 2018), 324.

20. Gröner, 1:168.

21. Lambi, 218, 222; "Das Torpedoboot," 960–61; Firle, "Die japanischen Torpedobootsunternehmungen im Kriege gegen Rußland: Erfolge und Lehren," *Marine Rundschau* 23, no. 12 (December 1912): 1585–86.

22. *Torpedobootstaktik*, 73–74, 97–98.

23. "Das Torpedoboot," 949–50, 951, 957.

24. ADM 186/17, *German Naval Warfare, Scouting and Guard Duties*, January 1915, 39.

25. "Das Torpedoboot," 952; Ernst Gagern, *Der Krieg in der Ostsee*, vol. 3 (Frankfurt: E. S. Mittler, 1964), 34; ONI, *Monthly Intelligence Bulletin*, 15 January 1921, 142.

26. Gröner, 1:173–74; Robert Gardiner, ed., *Conway's All the World's Fighting Ships 1906–1921* (London: Conway Maritime Press, 1985), 71–72.

27. Gröner, 1:176; Gardiner, ed., *Conway's 1906–1921*, 167; ONI, *Monthly Information Bulletin*, 15 May 1921, 12.

28. For concerns about size affecting the ability to make surprise night attacks, see "Das Torpedoboot," 956–57, and "Stand und Ausslichen der Torpedowaffe," 1095–96.

29. Friedman, *Naval Weapons*, 334; Arthur Hezlet, *The Electron and Sea Power* (New York: Stein and Day, 1975), 121; ONI, *Monthly Intelligence Bulletin*, 15 January 1922, 26–27.

30. *Taktische Befehle*, 37.

31. ADM 186/17, *German Naval Warfare, Scouting and Guard Duties*, 19, 34–35; *Torpedobootstaktik*, 71–71, 78.

32. ADM 137/4323, *Instructions Regarding the Use of A.S.T. (Scouting Signal Code)*, September 20, 1913, 68–72; ADM 186/17, *Maneuvering Instructions for the Fleet*, 1914, 71–74.

33. *Torpedobootstaktik*, 43.

34. *Torpedobootstaktik*, 121; ADM 186/17, *Scouting and Guard Duties*, 39.

35. Raymond C. Watson, Jr., *Radar Origins Worldwide: History of Its Evolution in 13 Nations Through World War II* (Victoria, BC: Trafford, 2009), 243.

36. ADM 137/4316, *Regulations for Making Recognition Signals*, 14 September 1914, 5–6.

37. Brooks, 507; John Campbell, *Jutland: An Analysis of the Fighting* (New York: Lyons Press, 2000), 277–78; ADM 137/4316, preface.

38. ADM 137/4272, German Documents. *Bestimmungen über die Abgabe von Erkennungssignalen* (Berlin: Admiralstab der Marine, 1916), 1 [hereinafter *Erkennungssignalen*].

39. *Erkennungssignalen*, 9.

40. *Torpedobootstaktik*, Tafel 5.

41. "RaMMD," *Marine Rundschau* 21, no. 3 (March 1910): 358; "RaMMD," *Marine Rundschau* 21, no. 6 (June 1910): 766.

42. "RaMMD," *Marine Rundschau* 22, no. 10 (October 1911): 1, 285.

43. *Rangliste der Kaiserlich Deutschen Marine für das Jahr 1914* (Berlin: Ernst Siegfried Mittler, n.d.), 60–66.

44. Groos, 1:Tabelle I.

45. *Torpedobootstaktik*, 48–49.

46. Reginald Belknap, "The Torpedo and Submarine Branches of the German Navy," *Naval Institute Proceedings* 42, no. 5 (September–October 1916): 1496–1500; Von Külhwetter, "The Personnel of the German Navy," in *The Naval Annual, 1913*, ed. Viscount Hythe (Portsmouth, UK: J. Griffin and Co.), 132–50.

47. Belknap, 1498, 1502.

48. *Taktische Befehle*, 63–66, 88–114, 175–178, 180–189.

49. "RaMMD," *Marine Rundschau* 21, no. 5 (May 1910): 631; "RaMMD," *Marine Rundschau* 21, no. 9 (September 1910): 1154; "RaMMD," *Marine Rundschau* 22, no. 7 (July 1911): 902; "RaMMD," *Marine Rundschau* 23, no. 8 (August 1912): 1126; "RaMMD," *Marine Rundschau* 24, no. 4 (April 1913): 489.

50. *Torpedobootstaktik*, 13; ONI, *Torpedo Boat Tactics, German Navy* (June 24, 1912), 1–2.

51. ADM 186/17, *Scouting and Guard Duties*, 33.

52. *Torpedobootstaktik*, 77, 98, 115–20.

53. ADM 186/17, *Scouting and Guard Duties*, 19, 35; *Taktische Befehle*, 13.

54. *Torpedobootstaktik*, 78–79.

55. Waldener, "Moderne Torpedoboote und ihre Artilleristische Bekämpfung," *Marine Rundschau* 17, no. 8 (August 1906): 929.

56. *Torpedobootstaktik*, 73.

57. *Torpedobootstaktik*, 47.

58. Lambi, 353–56.

59. Dirk Bönker, *Militarism in a Global Age: Naval Ambitions in Germany and the United States before World War I* (Ithaca, NY: Cornell University Press, 2016), 145.

60. *Torpedobootstaktik*, 47, 73.

61. Captain William S. Sims to Rear Admiral Bradley A. Fiske, April 2, 1915, 6.

62. Sims to Fiske, March 22, 1915, 5.

63. *Taktische Befehle*, 42–61.

64. Norman Friedman, *Fighting the Great War at Sea: Strategy, Tactics and Technology* (Annapolis: Naval Institute Press, 2014), 364n53.

65. Anonymous, "Strategical Principles and the Forth-Clyde Canal," *The Naval Review* 1, no. 4 (1913), 233–34.

66. Quoted in Lambi, 366.

67. Groos, 1:54.

68. Groos, 1:99–101.

69. Waldemar Vollerthun, *Der Kampf um Tsingtau: Eine Episode aus dem Weltkrieg 1914/1918* (Leipzig: S. Hirzel, 1920), 132–34.

70. ADM 186/17, *Scouting and Guard Duties*, 1.

71. Heinrich Rollmann, *Der Krieg in der Ostsee*, vol. 2 (Berlin: E. S. Mittler, 1929), 263, 265–67; N. B. Pavlovich, ed., *The Fleet in World War One*, vol. 1 (New Delhi: Amerind Publishing, 1979), 151–53; H. Graf, *The Russian Navy War and Revolution From 1914 up to 1918* (Honolulu: HI, University Press of the Pacific, 2002), 55–56; Vincent O'Hara and Leonard Heinz, *Clash of Fleets: Naval Battles of the Great War, 1914–1918* (Annapolis: Naval Institute Press, 2017), 130–31.

72. Details of service for German officers are drawn from the annual *Ranglisten der Kaiserlich Deutschen Marine* and their *Nachträge* published by Ernst Siegfried Mittler of Berlin. Additional details on Heinrich Schuur's career were kindly made available by his grandson, Konteradmiral Heinrich Schuur (Ret.) of the German Navy.

73. Groos, 4:270–74; *Monographs*, 14:146–54.

74. Groos, 5: 27–29; Reinhard Scheer, *Germany's High Sea Fleet in the World War* (London: Cassel, 1920), 102–103.

75. Groos, 5:35; Scheer, 107.

76. TNA, ADM 53/33420, *"Alyssum." Log February 1916*; ADM 53/36491, *"Buttercup." Log February 1916*; ADM 53/55501, *"Poppy." Log Month of February 1916*. The track chart in the German official history has the sloops looping back to the north after Boest's second attack, but they would have to steam at a speed of more than fifty knots to maneuver as the chart shows. Groos, 5:Karte B.

77. The chart in the official German history shows a total of four torpedoes being fired but the narrative says that the other two boats had muffed their targeting solutions and did not fire. Groos, 5:39, Karte B.

78. Groos, 5:35–43, Karte B; *Monographs*, 15:73–79.

79. *Nachtrag zur Rangliste der Kaiserlich Deutschen Marine für das Jahr 1912* (Berlin: Ernst Siegfried Mittler, n.d.), 60; *Rangliste für 1914*, 121, 127, 129.

80. Groos, 5:40.

81. Firle, "Japanischen Torpedobootsunternehmungen," 1577–78, 1582, 1588–89.

82. Groos, 5:38, 41–42, 44.

83. Quote: Groos, 5:44. In fairness, the design featured a triple bottom forward and cross and side bunkers to control flooding. D. K. Brown, *The Grand Fleet: Warship Design and Development, 1906–1922* (Annapolis: Naval Institute Press, 1999), 137.

84. Brooks, 506.

85. Brooks, 324; Campbell, 402.

86. Brooks, 401, 509–11; Campbell, 401–404.

87. O'Hara and Heinz, 191–94; Gagern, 3:34–35.

88. Friedman, *Naval Weapons*, 145–47.

89. Walter Gladisch, *Der Krieg in Der Nordsee*, vol. 6 (Berlin: E. S. Mittler, 1937), 194; Gröner, 1:178, 180; ADM 137/2085, "Destroyers," 1, 5; Henry Newbolt, *Naval Operations: History of the Great War*, vol. 4 (London: Longmans, Green: 1928), 52n2. Four of the boats assigned to the Dover Strait operation certainly carried 10.5-cm guns. Six (all those in 18th Half Flotilla and one in 17th Half Flotilla) certainly did not. The remaining fourteen likely received the heavier artillery sometime in the second half of 1916. The account of this operation is based on Gladisch, 6:147–49, 218–231, Karte 15; *Monographs*, 6:66–87, 17:185–189; Newbolt, 4:52–64; Reginald Bacon, *The Dover Patrol, 1915–1917*, vol. 2 (New York: Doran, 1919), 338–43, 435–36; Scheer, 187–89; Auguste Thomazi, *La Guerre Naval Dans la Zone des Armées du Nord* (Paris: Payot, 1924), 133–34; ADM 137/2085, "Memorandum of R. H. Bacon, Vice-Admiral, Dover," 3 November 1916.

90. Bacon, 1:49–51, 224, 292–93.

91. *Monographs*, 18:395.

92. Gerhard Groß, *Der Krieg in der Nordsee* (Hamburg: E. S. Mittler, 2006), 7:418.

93. David Morgan-Owen, *The Fear of Invasion: Strategy, Politics, and British War Planning, 1880–1914* (Oxford: Oxford University Press, 2017), 203.

94. ADM 137/3887, *German Submarine Silhouette Books: Torpedo Firing from Submarines*, January 1918, 3; Admiral George Callaghan, "Remarks on the Employment of Destroyers in a Fleet Action" (March 18, 1914), 2.

95. ADM 137/3887, 2, 59–64.

CHAPTER 3. THE BRITISH AND NIGHT FIGHTING

1. I am particularly grateful to Dr. Joe Moretz and to Peter Cannon for making available material from the United Kingdom's National Archives, which filled many gaps in my own research holdings. At a time when international travel and access to key archives are so restricted, this made a great difference. I am also grateful to Dr. Moretz and to Cdr. Andrew Livsey, RN, for very helpful criticism of my drafts.

2. Lt. Cdr. (later Rear Admiral) E. R. Corson, navigator of HMS *Caroline*, "Notes written on 1 June 1916 on board HMS *Caroline* of the 4th Light Cruiser Squadron after the Battle of Jutland," P. F. R. Corson, *Call the Middle Watch: An Account of Life at Sea in the Royal Navy 1905 to 1963* (Edinburgh: Pentland Press, 1997), 68.

3. Surgeon Capt. R. S. Allison, *H.M.S. Caroline* (Belfast: Blackstaff Press, 1974), 60–67.

4. TNA ADM 186/80, Tactical Division, Naval Staff, *O.U. 1683: Naval Tactical Notes* (Vol. I, 1929, Admiralty), 44.

5. "Remarks on the conduct of a fleet in action, based on the experience gained in the manoeuvres and exercises of the Home Fleets during the year 1913," H. F. Memorandum 0235 dated 5 December 1913. Box 1, Backhouse Papers, Naval Historical Branch.

6. ADM 137/289. *Grand Fleet Battle Orders* (*GFBO*) issued December 1915, 19.

7. *GFBO*.

8. "Addition to Battle Cruiser Orders: Para. 38: Lessons Learned from action of 31 May 1916," 31 August 1916. DRAX 1/3, Drax Papers, Churchill College. Reproduced in B. McL. Ranft (Ed.), *The Beatty Papers*, Vol. I, 1902–1918 (Aldershot, UK: Scolar Press, 1989), 372.

9. ADM 116/2090, HMS *Emperor of India* "Action Plotting," Memorandum Response dated 12 May 1921." *Emperor of India* had been one of the leaders in the development of tactical

plotting in 1916–18, which confirms concerns of the time about vital knowledge being lost without trace.

10. An officer of HMS *Broke* (anonymously), "Action off Jutland, May 31st, 1916" (*The Naval Review*, Vol. V, 1917), 174.

11. ADM 1/9266, "Statement of Dimensions HMS *Queen Elizabeth* following reconstruction."

12. G. Hermon Gill, *Royal Australian Navy 1942–1945* (Canberra: Australian War Memorial, 1968), 526.

13. N. J. M. Campbell, *Naval Weapons of World War Two* (London: Conway Maritime Press, 1985), 5.

14. Norman Friedman, *Naval Firepower: Battleship Guns and Gunnery in the Dreadnought Era* (Barnsley, UK: Seaforth), 155.

15. ADM 137/1645, "Post-Jutland Changes:" Home Fleets letter No. 1483/H.F. 1187 of 29 June 1916.

16. Adm. Sir Studholme Brownrigg, "Gunnery in the Royal Navy: Firing at Night," Adm. Sir Reginald Bacon (Ed.), *Britain's Glorious Navy* (London: Odhams, 1943), 222.

17. ADM 186/339, Training and Staff Duties Division, Naval Staff, *CB 3001/1914–36. Summary of Progress in Naval Gunnery, 1914–1936* (Admiralty, December 1936), 85.

18. ADM 137/1645, "Admiralty letter M06016 of 4 July 1916."

19. Ranft, 1:364.

20. Training and Staff Duties Division, Naval Staff, *CB 917Q: Naval Staff Monographs (Historical) Vol. XVIII, Home Waters—Part VIII: December 1916—April 1917* (Admiralty, May 1933), 274 [hereinafter *Monographs*]. The Naval Staff Monographs can be downloaded from the Royal Australian Navy's Sea Power Centre—Australia (SPC-A) site at: https://www.navy.gov.au/media-room/publications/wwi-naval-staff-monographs.

21. James Goldrick, *After Jutland: The Naval War in Northern European Waters, June 1916—November 1918* (Annapolis: Naval Institute Press, 2018), 237.

22. James Goldrick, "Work Up," in Stephen Howarth and Derek Law (Eds.), *The Battle of the Atlantic 1939–1945: The 50th Anniversary International Conference* (London: Greenhill Books, 1994), 220–39.

23. *Monographs*, 278.

24. Goldrick, *After Jutland*, 219.

25. *Monographs*, 399–401.

26. Goldrick, *After Jutland*, p. 154.

27. Tactical Division, Naval Staff, *CB 1769/28 (2) Selected Reports of Exercises, Operations and Torpedo Practices in H.M. Fleet (Vol. II) Summer and Autumn 1927* (Admiralty, September 1928), 26.

28. Tactical Division, Naval Staff, *CB 1769/30 (1) Exercises and Operations 1930* (Vol. I), Admiralty, 1930, 3; Tactical Division, Naval Staff, *CB 1769/29 (2) Exercises and Operations 1929* (Vol. II) (Admiralty, June 1930), 9.

29. Christina J. M. Goulter, *A Forgotten Offensive: Royal Air Force Coastal Command's Anti-Shipping Campaign, 1940–1945* (London: Frank Cass, 1995), 59.

30. *CB 1769/28 (2) Selected Reports*, 10.

31. Tactical Division, Naval Staff, *CB 3016/34: Progress in Tactics 1934 Edition* (Admiralty, August 1934), 67, 68.

32. Adm. Sir William Fisher letter to Rear Adm. Max Horton, 25 December 1935. Cited in W. S. Chalmers, *Max Horton and the Western Approaches* (London: Hodder & Stoughton, 1954), 39.

33. *CB 1769/28 (2) Selected Reports; Progress in Tactics 1928* (Admiralty, August 1929), 8.

34. *CB 3016/34: Progress in Tactics 1934*, 67.

35. Pound to Chatfield letter dated 31 March 1936, in Paul G. Halpern (Ed.), *The Mediterranean Fleet 1930–1939* (Abingdon, UK: Routledge for The Navy Records Society, 2016), 161.

36. F. A. Kingsley (Ed.), *The Development of Radar Equipments for the Royal Navy, 1935–45* (London: Macmillan, London, 1995), xxv.

37. ADM 186/451, Tactical Division, Naval Staff, *CB 3002/37 Progress in Torpedo, Mining, Minesweeping, A/S Measures, and Chemical Warfare Defence 1937 Edition* (Admiralty, October 1937), 8.

38. Torpedo Division, Tactical Section, Naval Staff, *CB 1769/27 (2) Selected Reports of Exercises, Operations and Torpedo Practices in H.M. Fleet Summer and Autumn 1927* (Admiralty, September 1928), 26.

39. http://www.arl-teddington.org.uk/arl_history_1921–44.htm.

40. Admiralty Fleet Order 2989/28 *Brewerton Table*, SPC-A.

41. ADM 178/223, Cdr. C. R. L. Parry RN Letter of Proceedings No. 9, 10 March 1941.

42. Admiralty Fleet Order 3089/26 *Course Records-Brewerton Type—Supply*, SPC-A.

43. Admiralty Fleet Order 2642/26 *Strategical and Tactical Plotting*, SPC-A.

44. Drax's initial letter was written to the Director of the Training and Staff Duties Division in January 1920. See ADM 116/2090.

45. Confidential Admiralty Fleet Order 1820/23. The responses are contained in ADM 1/8662/109 "Action Plotting Arrangements: Standardisation of Battleships and Cruisers."

46. https://collections.rmg.co.uk/collections/objects/42264.html.

47. The ship would be transferred to the RAN in 1943 to replace the sunken *Canberra*.

48. W. S. Chalmers, *Max Horton and the Western Approaches* (London: Hodder & Stoughton, 1954), 44.

49. As was the destroyer *Obdurate's* experience after the Battle of Jutland. H. P. K. Oram, *Ready for Sea* (London: Seeley Service, 1974), 167–68.

50. Tactical Division, Naval Staff, *CB 1769/28 (2) Selected Reports of Exercises, Operations and Torpedo Practices in H.M. Fleet* (Vol. II) *and Progress in Tactics*, 20.

51. Vice Adm. Sir James Troup letter to Captain S. W. Roskill dated 4 July 1964. Cited in Stephen Roskill, *Naval Policy between the Wars*, Vol. 1, *The Period of Anglo-American Antagonism 1919–1929* (London: Collins, 1968), 533n3.

52. Training and Staff Duties Division, Naval Staff *CB 973: The Naval War Manual, 1925* (Admiralty, October 1925), 31. Copy held by author.

53. Admiral of the Fleet Lord Chatfield, *The Navy and Defence* (London: William Heinemann, 1942), 240.

54. See Joseph Moretz, *The Royal Navy and the Capital Ship in the Interwar Period: An Operational Perspective* (London: Frank Cass, 2002), 226–27. Also, Andrew Field, *Royal Navy Strategy in the Far East 1919–1939: Preparing for War against Japan* (London: Frank Cass, 2004), 176–77.

55. Adm. Sir William James, "Night Fighting at Sea by the British Navy," U.S. Naval Institute *Proceedings* (Vol. 70, No. 6, June 1944), 645–61.

56. Lt. Cdr. Colin A. G. Hutchison (anonymously), "Battle by Night," *The Naval Review* (Vol. X, No. 1, February 1922), 14.

57. Jon Tetsuro Sumida, "The Best Laid Plans: The Development of British Battle-Fleet Tactics, 1919–1942," *The International History Review* (Vol. 14, No. 4, November 1992), 688.

58. Rear Adm. Reginald Plunkett-Ernle-Erle-Drax (anonymously) "Battle Tactics," *The Naval Review* (Vol. XX, No. 4, November 1932), 656.

59. Capt. S. W. C. Pack, *Cunningham the Commander* (London: B.T. Batsford, 1974), 62.

60. A. B. Cunningham, *A Sailor's Odyssey: The Autobiography of Admiral of the Fleet Viscount Cunningham of Hyndhope K.T G.C.B D.S.O* (London: Hutchinson, 1951), 161–62.

61. Lt. Cdr. J. H. Walker, RAN, in a contemporary letter, cited in Paul McGuire & Frances McGuire, *The Price of Admiralty* (Melbourne: Oxford University Press, 1944), 98.

62. Adm. Sir Studholme Brownrigg, "Gunnery in the Royal Navy," Adm. Sir Reginald Bacon (Ed.), *Britain's Glorious Navy* (London: Odhams, 1943), 221.

63. Joseph Moretz, *Thinking Wisely, Planning Boldly: The Higher Education and Training of Royal Navy Officers, 1919–1939* (Solihull, UK: Helion, 2014), 234.

64. Vice Adm. Howard Kelly to Adm. Sir Roger Keyes, letter dated 14 April 1936, Paul G. Halpern (Ed.), *The Keyes Papers*, Vol. II, *1919–1938* (London: George Allen and Unwin, 1980), 177.

65. Captain E. S. Brand, RN, letter to J. Dixon 26 November 1976. In possession of author. See also Geoffrey Lowis, *Fabulous Admirals and Some Naval Fragments* (London: Putnams, 1957), 243–44.

66. ADM 186/304, Gunnery Division, Naval Staff, Admiralty *CB 3001/30 Progress in Gunnery, 1930.*

67. Lt. Cdr. J. H. Walker, RAN, in a contemporary letter, cited in McGuire and McGuire, 99.

68. ADM 186/154, Tactical Division, Naval Staff, *CB 1769/33 (II) Exercises and Operations, 1933* (Admiralty, April 1934), 2.

69. The narrative of Exercise Z.H. derives from *CB 1769/34 (I) Exercises and Operations, 1934.* (Admiralty, November 1934), as well accounts by those involved. Eric Brand, who as staff officer for operations to the commander in chief Home Fleet was charged with the post-exercise analysis, had to admit that "after the battleships had met in the night I gave up trying to sort out or describe the melee during which nobody had taken any records." Capt. E. S. Brand, RN, letter to J. Dixon, 26 November 1976.

70. Capt. E. S. Brand, RN, letter to J. Dixon, 26 November 1976; "A War between Redland and Blueland: Manoeuvres on Grand Scale," *Evening News* (Rockhampton, Queensland, 9 March 1934), 5. Accessed at: https://trove.nla.gov.au/newspaper/article/198725664?searchTerm =fleet%20manoeuvres%20admiral%20fisher%20boyle.

71. Adm. Sir William James, *The Sky Was Always Blue* (London: Methuen, 1951), 177.

72. Soon to succeed as Earl of Cork and Orrery.

73. William Henry Dudley Boyle Cork and Orrery (Earl of), *My Naval Life* (London: Hutchinson, 1942), 165.

74. Adm. Sir William James, *Admiral Sir William Fisher* (London: Macmillan, 1943), 127.

75. James, 127.

76. Naval Staff, *Battle Instructions 1934*, Admiralty, 1934. Cited in Norman Friedman, *Naval Firepower: Battleship Guns and Gunnery in the Dreadnought Era* (Barnsley, UK: Seaforth), 114.

77. Gino Jori, "Pensiero strategico navale e costruzioni navali," *Rivista Marittima* (No. 2, 1989), 104.

78. Enrico Cernuschi, "*Marinelettro e il Radiotelemetro Italiano: Lo sviluppo e l'evoluzione del radar navale (1933–1943)*" (*Supplemento al Rivista Marittima, May 1995*), 25–26.

79. Tactical Division, Naval Staff, *CB 3016/35: Progress in Tactics 1935 Edition* (Admiralty, October 1935), 82.

80. H. P. Willmott, *Empires in the Balance: Japanese and Allied Pacific Strategies to April 1942* (London: Orbis, 1982), 354.

81. John Winton, *Carrier Glorious: The Life and Death of an Aircraft Carrier* (London: Cassell, London 1999), 25. The pilot was Lt. (later Cdr.) H. P. Madden in a Fairey IIIF.

82. Winton, 43.

83. Tactical Division, Naval Staff, Admiralty, *CB 1769/31 (1) Exercises and Operations 1931.*

84. Tactical Division, Naval Staff, Admiralty, *CB 03016/39: Progress in Tactics 1939 Edition* (Admiralty, June 1939), 16.

85. Fisher to Chatfield letter dated 25 February 1936, in Halpern, *Mediterranean Fleet*, 154.

86. *CB 03016/39: Progress in Tactics 1939*, 15.

87. Norman Friedman, *British Carrier Aviation: The Evolution of the Ships and their Aircraft* (Annapolis: Naval Institute Press, 1988), 165.

88. *CB 03016/39: Progress in Tactics 1939*, 22.

89. *CB 3016/35: Progress in Tactics 1935*, 83.

90. ADM 186/451, Tactical Division, Naval Staff, *CB 3002/37 Progress in Torpedo, Mining, Minesweeping, A/S Measures, and Chemical Warfare Defence 1937 Edition* (Admiralty, October 1937), 16.

91. Tactical Division, Naval Staff, "R.A.F. Exercises," *CB 1769/34 (I) Exercises and Operations, 1934* (Admiralty, November 1934), 82–85.

92. A. C. Wright, "The Royal Air Force in the Far East," *Royal United Services Institute Journal,* 80: 531–32, 536.

93. UK TNA, AIR 23/7761 "Royal Air Force Far East—Monthly General Summaries of Work: January 1937—July 1939."

94. Vincent Orange, *Tedder: Quietly in Command* (London: Frank Cass, 2004), 99–100.

95. ADM 116/3121, C-in-C China N053/01526, "Report of the Singapore Conference January 1934" and covering Naval Staff comments. Australian Joint Copying Project Microfilm, National Library of Australia. See also ADM 116/3862 "Naval Arrangements in the Event of War in the Far East" Director of Plans Minute dated 24 July 1936. The exchange between the conference delegates and the local RAF commander is remarkable for the accuracy of the 1934 naval assessment of Japanese carrier capability—suggesting that a mass strike of a hundred aircraft could be launched up to two hundred miles from Singapore and that the Japanese would do so at dawn after a night approach. The delegates were citing the British naval attaché in Tokyo, who seems at this time to have been extremely well-informed.

96. Termed Asdic by the British for many years.

97. Willem Hackmann, *Seek & Strike: Sonar, Anti-Submarine Warfare and the Royal Navy 1914–54* (London: HMSO, 1984), 220.

98. *CB 03016/39: Progress in Tactics 1939*, 56.

99. Little detail is provided for this figure, but the range suggests that it may have been a "convergence zone" detection, a phenomenon that would be of great significance in the effort to detect Soviet submarines during the Cold War.

100. ADM 186/522, Tactical Division, Naval Staff, *CB 3002/35 Progress in Torpedo, Mining,*

Minesweeping, A/S Measures, and Chemical Warfare Defence 1935 Edition (Admiralty, September 1935), 73.

101. *CB 03016/39: Progress in Tactics 1939*, 52–53.

102. Oliver Warner, *Cunningham of Hyndhope, Admiral of the Fleet* (London: John Murray, 1967), 79.

103. Patrick Beesly, *Very Special Admiral: The Life of Admiral John H. Godfrey, CB* (London: Hamish Hamilton, 1980), 82.

104. Cunningham, 336.

CHAPTER 4. FORCED TO FIGHT

1. Archivio dell'Ufficio Storico della Marina Militare (AUSMM), Comando R. Torp, *Lupo*, "Rapporto di missione," 26 May 1941.

2. Frank Wade, *Midshipman's War: A Young Man in the Mediterranean Naval War 1941–1943* (Victoria, Canada: Trafford, 2005), 55.

3. Andrew Browne Cunningham, *A Sailor's Odyssey* (London: Hutchinson & Co., 1951), 369.

4. Guido Candiani, *I vascelli della Serenissima* (Venice: Istituto Veneto di Lettere, Scienze e Arti, 2009), 539.

5. Mario Navi Mocenigo, *Storia della Marina veneziana da Lepanto alla caduta della Repubblica* (Venice: Filippi Editore, 1995), 333; Candiani, *Serenissima*, 541–42.

6. Guglielmo Marconi, the creator of the first practical wireless, supervised a successful test in 1897 and the Italian navy began adopting radio the next year.

7. D. Bonamico, "Il governo tattico delle flotte," *Rivista Marittima* (July 1902), 249.

8. AUSMM, *Norme e direttive da emanare per l'impiego delle forze in caso di conflitto*, Appendice N.1.

9. Ezio Ferrante, "Il Grande ammiraglio Paolo Thaon di Revel," *Supplemento della Rivista Marittima* (August–September 1989), 47, 63.

10. Monthly merchant departures to Albania averaged seventy-two throughout the course of the war. Achille Rastelli and Alessandro Massignani, ed., *La guerra navale 1914–1918* (Valdagno, Italy: Gino Rossato Editore, 2002), 325.

11. Ferrante, 47, 63.

12. For this period see O'Hara and Heinz, 150–61, 200–209, 259–73, 282–288; Enrico Cernuschi and Vincent P. O'Hara, "The Naval War in the Adriatic Part 1," *Warship 2015* (London: Conway Maritime Press, 2015), 161–73; and "The Naval War in the Adriatic Part 2," *Warship 2016* (London: Conway Maritime Press, 2016), 62–75.

13. Ecole de guerre navale, "Le problème général du tir à la mer," Session 1924–25; Marc Saibène, *Le torpilleurs lègers Français 1937–1945* (Rennes: Marines Editions, 2004).

14. See John Gooch, *Mussolini's War: Fascist Italy from Triumph to Collapse 1935–1943* (New York: Pegasus, 2020).

15. For Italy's naval policy and strategy see Pier Paolo Ramoino, "Una storia 'Strategica' della Marina Militare Italiana," *Rivista Marittima* (*Supplemento della Rivista Marittima*, September, 2018).

16. Edoardo Vollono, "Il siluro un'arma nuova," *Rivista Marittima* (April 1961), 42.

17. Ettore Aymone, "Problema tecnico del lancio del siluro," *Rivista Marittima* (June 1938).

18. Erminio Bagnasco, *Le armi delle navi italiane nella Seconda Guerra Mondiale* (Parma: Albertelli, 2007), 45.

19. Fabio Tani, *Ricordanze, uomini, navi e cannoni* (Rome: Privately published, 2016), 97.

20. Ramoino, "Storia 'Strategica,'" 94–95.

21. Antonio Petroni, "Siluri, lanciasiluri e lanci," *Rivista Marittima* (November 1938), 381–82.

22. Giuseppe Fioravanzo, *La Marina Italiana nella seconda guerra mondiale*, Volume XXI *L'organizzazione della Marina durante il conflitto Tomo I* (Rome: Ufficio Storico della Marina Militare (USMM), 1972), 100.

23. See Giuseppe Finizio, "I binocoli della Regia Marina tra le due guerre mondiali," *Rivista Marittima* (July–August 2016).

24. See Franco Bargoni, "L'intervento navale italiano nella Guerra civile Spagnola," *RID Rivista Italiana Difesa* (March 1987).

25. See Francesco Mattesini, *Corrispondenza e direttive tecnico-operative di Supermarina*. I:2 (Rome: USMM, 2000), 769–70.

26. Mattesini, 769.

27. See "Relazione sulla operazione," 20 October 1940, in Mattesini, 769–73; ADM 199/797, *Ajax* LOP; Bragadin, 38.

28. ADM 199/797, 1–2.

29. Mattesini, 769.

30. AUSSM, *Le Azioni navale*, Notizia 1, "Azione notturna dell'11–12 Ottobre," 1:6.

31. Quotes: ADM 199/797, LOP, Enclosure 2, 1; Mattesini, 771.

32. *Azione navale*, 1:7.

33. ADM 199/797, LOP, 7.

34. ADM 199/797, "Remarks of Vice Admiral, Light Forces, on Enclosure No. 2," 2.

35. AUSSM, *Azione navale* 4, "Attacco delle torpediniere *Lupo* e *Libra* contro un convoglio nemico 30–31 gennaio," 3.

36. AUSMM, *Norme di massima per l'impiego in guerra*, January 1942.

37. AUSMM, Comando in Capo SN "Navigazione notturna in guerra," 22 April 1941.

38. AUSSM, "Comando Capo Squadra Navale a N. Murzi," 16 May 1941.

39. Erminio Bagnasco and Augusto de Toro, *The Littorio Class: Italy's Last and Largest Battleships 1937–1948* (Annapolis: Naval Institute Press, 2011), 101.

40. Luigi Carilio Castioni, "I radar industriali italiani," *Storia contemporanea* (Dicembre 1987), 1244–45.

41. Vincent P. O'Hara, *Struggle for the Middle Sea: The Great Navies at War in the Mediterranean Theater, 1940–1945* (Annapolis: Naval Institute Press, 2009), 62–64.

42. Aldo Cocchia, *La Marina Italiana nella seconda guerra mondiale*, Volume VII: *La difesa del traffico con l'Africa settentrionale dal 1° ottobre 1941 al 30 settembre 1942* (Rome: USMM, 1962), 150. Also see discussion in Vincent P. O'Hara, *Six Victories, North Africa, Malta, and the Mediterranean Convoy War November 1941–March 1942* (Annapolis: Naval Institute Press, 2019), 41–43.

43. See Vincent P. O'Hara, *German Fleet at War 1939–1945* (Annapolis: Naval Institute Press, 2004), 107–8.

44. ADM 199/897, *Aurora*, "Report of Proceedings (ROP), 8; O'Hara, *Six Victories*, 46.

45. Cocchia, 58, 64.

46. O'Hara, *Six Victories*, 48.

47. Christopher Page, ed., *The Royal Navy and the Mediterranean*, Volume II: *November 1940–December 1941*. (London: Frank Cass, 2002), 192.

48. See O'Hara, *Six Victories*, 44–53.

49. Quotes, Iachino, 49–50; Mattesini, II/2:1166. Also, USMM, *Difesa del traffico,* VII:515.

50. See O'Hara, *Six Victories*, 36.

51. Franco Bargoni and Franco Gay, *Corazzate classe "Caio Duilio"* (Rome: Bizzarri, 1972), 73.

52. AUSSM, Supermarina, "Situazione Nafta per Caldale," 21 February and 1 April 1942; O'Hara, *Six Victories*, 37.

53. Many other factors were at play, not least the increased bombardment of Malta by German air forces, but the run of successful convoys started before the first German attacks against Malta.

54. Rolf Johannesson, *Integrierung im Mittelmeer 1942, Rivista Marittima* (June 1953), 508–11.

55. Bragadin, 239.

56. AUSMM, *Le Azioni navale,* Notizia 107, "Attacco di Unità di superficie nemiche ad un nostro convoglio nel Canale di Sicilia," 2 August 1943, 12.

57. Basil Jones, *And so to Battle: A Sailor's Story* (Privately Published, 1979), 67.

58. Convoys that stood off weaker attackers include two instances in the Red Sea.

59. O'Hara, *Middle Sea*, 216–17.

CHAPTER 5. HOW CAN THEY BE THAT GOOD?

1. See David Evans and Mark Peattie, *Kaigun: Strategy, Tactics and Technology in the Imperial Japanese Navy 1887–1941* (Annapolis: Naval Institute Press, 1997), 210–11 [hereinafter *Kaigun*]; Eric Lacroix and Linton Wells II, *Japanese Cruisers of the Pacific War* (Annapolis: Naval Institute Press, 1997), 184–85 [hereinafter JCPW]; TROM for *Jintsu* at http://www.combinedfleet.com/jintsu_t.htm.

2. *Kaigun*, 211.

3. Japan had signaled its unhappiness with and intent to withdraw from the Treaty system on 29 December 1934. The subsequent departure of Japan's delegation during the London Treaty negotiations on 15 January 1936 not only formalized the matter, but also effectively destroyed the larger Treaty structure by removing one of its most important signatories. Sadao Asada, *From Mahan to Pearl Harbor: The Imperial Japanese Navy and the United States* (Annapolis: Naval Institute Press, 2013), 198; *Kaigun*, 298.

4. *Kaigun*, 141–42.

5. *Kaigun*, 143–4; Asada, 74–79.

6. *Kaigun*, 129, 132, 201–4, 277–86.

7. *Kaigun*, 205–7, 250–66.

8. See Jeff Alexander, "Nikon and the Sponsorship of Japan's Optical Industry by the Imperial Japanese Navy, 1917–1945," *B.C. Asian Review* 13 (Spring 2002): 1–21; C. G. Grimes, ed. "Japanese Optics," in *U.S. Naval Technical Mission to Japan* [hereinafter NTMJ], Series 10: Miscellaneous Targets, Report X-05 (Washington: U.S. Government Printing Office, U.S. Naval History Division, 1945).

9. JCPW, 236, 772–73.

10. JCPW, 761.

11. JCPW, 235, 773.

12. JCPW, 235.

13. "Japanese Propellants–General," NTMJ, Report O-10-2, 5–7.

14. Schematics examples of the communications net for both the Pearl Harbor and Midway operations illustrating these principles can be found in U.S. Armed Forces Far East, History Division, Japanese Monograph 118, *Operational History of Japanese Naval Communications*, 238, 257 [on line https://cdm16040.contentdm.oclc.org/digital/collection /p4013coll8/id/2434] [hereinafter Monograph 118].

15. Monograph 118.

16. JCPW, 334.

17. A perusal of Lacroix and Wells reveals that during any major refit of a Japanese cruiser, its communications facilities were likely to be upgraded.

18. JCPW, 282.

19. Monograph 118, 251.

20. "Japanese Radio Communications and Radio Intelligence: 'Know your Enemy!,'" CinCPac - CinCPOA Bulletin 5–45. https://www.ibiblio.org/hyperwar/USN/ref/KYE/CINCPAC -5–45/index.html.

21. JCPW, 334.

22. See in particular Roger Wilkinson, "Short Survey of Japanese Radar—I," *Electrical Engineering*, Vol. 65 (Aug-Sep 1946), 370–77; Roger Wilkinson, "Short Survey of Japanese Radar—II," *Electrical Engineering*, Vol. 65 (Oct 1946), 455–63; "Japanese Radio, Radar, and Sonar Equipment," NTMJ Report E-17; *Kaigun*, 394, 411, 503, 505–8, 595; JCPW, 773–77.

23. JCPW, 774.

24. Wilkinson, 372.

25. *Kaigun*, 595; Monograph 118, 112.

26. JCPW, 333–4, 350, 575. These facilities were apparently installed on all surviving "A-class" [heavy] cruisers, as well as the modern light cruiser *Sakawa*.

27. A perusal of various "A-Class" cruiser plans in JCPW reveals no "information rooms" installed in 1942.

28. *Kaigun*, 46–47.

29. *Kaigun*, 223.

30. Frederick Milford, "Imperial Japanese Navy Torpedoes, Part II: Heavyweight Torpedoes 1918–1945," *The Submarine Review* (July 2002) 53–54; John Campbell, *Naval Weapons of World War Two* (Annapolis, MD: Naval Institute Press, 2002), 204.

31. Jiro Itani, Hans Lengerer, and Tomoko Rehm-Takahara, "Japanese Oxygen Torpedoes and Fire Control Systems," *Warship* 1991 (London: Conway's Maritime Press, 1991), 121–33; *Kaigun*, 266–70; JCPW, 780.

32. Milford, 58.

33. Itani, et al., 123; "Japanese Torpedoes and Tubes, Article 1: Ship and Kaiten Torpedoes," NTMJ, Report O-01–1, 84.

34. Milford, 59; Itani, et al., 123–24.

35. Sources on this matter are conflicting. *Kaigun* cites Kishimoto as the team lead, whereas Itani names Oyagi as the technical lead. NTMJ refers to Oyagi as an important contributor (he was interviewed by the Americans after the war), while Oyagi cites his own efforts as well as several other individuals. It seems clear that the Japanese deployed a large, multidisciplinary team to overcome the technical obstacles in their way.

36. *Kaigun*, 267.

37. Milford, 60; Campbell, 207.

38. NTMJ, 1, 148, 152.

39. NTMJ, 240.

40. "Action Report—Night Engagement off KOLOMBANGARA during night of 12–13 July 1943," Commander Task Force Eighteen, Serial 00118, 3 August 1943, 9. I am indebted to Vince O'Hara for his help in this matter.

41. "Japanese Torpedoes and Tubes, Article 3, Above-Water Tubes," NTMJ, Report O-01-3, 17–19; JCPW, 248.

42. JCPW, 136–7.

43. "Japanese Torpedo Fire Control," NTMJ, Report O-32, 1; Itani, et al., 128–31.

44. Itani, et al.

45. It is worth noting with respect to torpedo carrying surface craft that the Japanese never employed motor torpedo boats with the same enthusiasm that other combatants such as the U.S., British, Italians, or Germans did. Although they did construct a number of vessels later in the war based on German Schnellboot plans, information on them is scanty. See Hansgeorg Jentschura, Dieter Jung, and Peter Mickel, *Warships of the Imperial Japanese Navy, 1869–1945* (Annapolis: Naval Institute Press, 1986), 154–58. While speculative, it is possible that this may have been due to the Imperial Navy's focus on Decisive Battle. Because this was primarily a blue-water mission, it may have led to a lack of interest in littoral warfare.

46. *Kaigun*, 221–22; M. J. Whitley, *Destroyers of World War Two* (Annapolis: Naval Institute Press, 1988), 192–93.

47. In practice, however, the maximum elevation of 55 degrees, coupled with the slow traverse of the turrets, meant that they were unable to be used as true dual-purpose weapons.

48. M. J. Whitley, *Destroyers*, 193.

49. *Kaigun*, 222.

50. "Japanese Sonar and Asdic," NTJM, Report E-10, 11, 21.

51. *Myōkō* originally was armed with 20-cm (7.9 inch) weapons, but these were later upgraded to true 20.3-cm (8-inch) guns.

52. Whitley, *Cruisers of World War Two* (Annapolis: Naval Institute Press, 1995), 174–75; JCPW, 264.

53. JCPW, 64–5, 83–4.

54. JCPW, 221–2.

55. JCPW, 222.

56. JCPW, 466, 515.

57. JCPW, 41, 55–6.

58. JCPW, 88–92.

59. JCPW, 137–38.

60. *Kaigun*, 220–21.

61. *Kaigun*, 223.

62. *Kaigun*, 276.

63. *Kaigun*, 277. Initially only *Kirishima* and *Haruna* had been rebuilt and were capable of this role. It should also be noted that while BatDiv 4 is mentioned in Japanese source materials, it was never actually formed. The four fast *Kongō* units all remained assigned to BatDiv 3 during the Pacific War.

64. *Kaigun*, 280. It should be noted that the Germans had used battleships in support of mining operations in the North Sea during World War I. Likewise, American doctrine at the time was also moving to a position where cruisers supported destroyers when involved in Night Search and Attack.

65. *Kaigun*, 277–78.

66. *Kaigun*, 280.

67. Richard Frank, *Guadalcanal: The Definitive Account of the Landmark Battle* (New York: Penguin, 1990), 85.

68. *Kaigun*, 211; Whitley, *Destroyers*, 193.

69. *Kaigun*, 242–43; JCPW, 719–21.

70. JCPW, 721–24; *Kaigun*, 243.

71. *Kaigun*, 244–46.

72. Paul Dull, *A Battle History of the Imperial Japanese Navy, 1941–1945* (Annapolis: Naval Institute Press, 1982), 73–88; Japanese National Institute for Defense Studies, *Bōeichō Bōeikenshūjō Senshibu* (originally, *Bōeichō Boeikenshūjō Senshishitsu*), also *Senshi Sōsho* (War history) series, *The Operations of the Navy in the Dutch East Indies and the Bay of Bengal*, Vol. 26, 1969, trans. by Willem Remmelink (Leiden, Netherlands: Leiden University Press, 2018), 431–63, 472–74 [hereinafter *Senshi Sōsho*]; Donald Kehn, *In the Highest Degree Tragic: The Sacrifice of the U.S. Asiatic Fleet in the East Indies during World War II* (Dulles, VA: Potomac Books, 2017), 289–333.

73. JCPW, 212; Campbell, 204–5; Whitley, *Destroyers*, 193, 196–97; Richard Worth, *Fleets of World War II* (Cambridge: DaCapo, 2001), 195. Not all of the Japanese torpedoes were Type 93—the two *Fubuki* class carried Type 90.

74. Kehn, 294.

75. *Senshi Sōsho*, 453.

76. Kehn, 318–20.

77. Kehn, 328.

78. *Senshi Sōsho*, 454–55.

79. *Senshi Sōsho*, 473–74.

80. *Senshi Sōsho*, 452.

81. "Detailed engagement report from August 8 to 9, 1942 on Battle of Savo Island, Warship *Chokai* (Night Battle of Tulagi Strait)," JACAR (www.jacar.archives.go.jp), C08030747100; Trent Hone, *Learning War: The Evolution of Fighting Doctrine in the U.S. Navy, 1989–1945* (Annapolis: Naval Institute Press, 2018), 174–79; James D. Hornfischer, *Neptune's Inferno: The U.S. Navy at Guadalcanal* (New York: Random House, 2011), 78–85; Frank, 83–123; Dennis Warner, Peggy Warner, and Sadao Seno, *Disaster in the Pacific: New Light on the Battle of Savo Island* (Annapolis: Naval Institute Press, 1992); Dull, 184–95; Commodore Richard W. Bates, USN (Ret.) and Commander Walter D. Innis, USN, *The Battle of Savo Island, August 9th, 1942: Strategical and Tactical Analysis* (Newport, RI: U.S. Naval War College, 1950) [hereinafter Analysis].

82. Turner's later attempts to place blame on the Australians for faulty sighting were clearly dishonest (Warner, et al., 4–5, 72–78), as were many of his attempts to deflect blame for the disaster.

83. Analysis, 115; *Chōkai* Battle Report track chart; Frank, 341.

84. If *Canberra* was hit by a torpedo, it seems most probable that it would have been inflicted

by USS *Bagley*. See in particular Bruce Loxton with Chris Coulthard-Clark, *The Shame of Savo: Anatomy of a Naval Disaster* (Annapolis: Naval Institute Press), 182–206. I thank James Goldrick for bringing this source to my attention.

85. Analysis, 115.

86. Warner, et al., 212.

87. Warner, et al., 258.

88. Frank, 292–312; Dull, 215–21; Hone, 185–89; Hornfischer, 154–85.

89. Frank, 299.

90. Dull, 237–42; Frank, 428–61; Hornfischer, 253–323; Hone, 190–7.

91. Hone 192–94.

92. Frank, 441.

93. Dull, 243–47; Hone, 198–201; Hornfischer, 332–66; Robert Lundgren, "Kirishima Damage Analysis," September 2010, and "The Battleship Action 14–15 November 1942," March 2010 (http://www.navweaps.com/index_lundgren/Biography_Robert_Lundgren.php); "Action Report, Night of November 12–15, 1942," in John C. Reilly Jr., ed., *Operational Experience of Fast Battleships: World War II, Korea, Vietnam* (Washington: Naval Historical Center, 1989), 61–67.

94. Hornfischer, 362.

95. Most accounts credit *Washington* with between six and nine large-caliber hits, but recent scholarship indicates a much higher total. Lundgren, "Battleship Action," 21–22.

96. Dull, 255–58; Frank, 493–518; Hone, 203–5.

97. Frank, 516.

98. Frank, 518.

99. Hone, 197.

100. "Action Report—Night Engagement off KOLOMBANGARA during night of 12–13 July 1943," Commander Task Force Eighteen, 9.

101. Their slower top speed and heavy fuel needs may have influenced this decision.

102. Ugaki Matome, *Fading Victory The Diary of Admiral Matome Ugaki 1941–1945*, ed. by Donald M. Goldstein and Katherine V. Dillon, trans. by Masataka Chihaya (Pittsburgh: University of Pittsburg Press, 1991), 299.

103. Ugaki, 299.

104. Based on figures from David Brown, *Warship Losses of World War Two* (Annapolis, MD: Naval Institute Press, 1995). Japanese submarine losses developed from http://www.ibiblio.org/hyperwar/Japan/IJN/JANAC-Losses/JANAC-Losses-3.html, and checked against individual TROMs on www.combinedfleet.com. Including early-war British and Dutch losses would, of course, improve Japan's overall ratio.

105. Hornfischer, 235.

CHAPTER 6. MASTERING THE MASTERS

1. Letter from William S. Sims to Rear Adm. Bradley A. Fiske, 22 March 1915, Naval War College Archives, Record Group 8, Box 103.

2. Sims, Letter, 22 March 1915.

3. Sims.

4. Lt. Cdr. Dudley W. Knox, "The Role of Doctrine in Naval Warfare," *Proceedings* 41, no. 2 (1915): 325–54.

5. Elting E. Morison, *Admiral Sims and the Modern American Navy* (Boston: Houghton Mifflin Company, 1942), 295.

6. Knox.

7. Trent Hone, *Learning War: The Evolution of Fighting Doctrine in the U.S. Navy, 1898–1945* (Annapolis: Naval Institute Press, 2018), 116; Trent Hone, "Exploring the Options—The Development of USN Tactical Doctrine, 1913–23," *Naval War College Review* 72, no. 4 (2019): 101–23.

8. "Comment by Commander Battleship Division Four," in *United States Fleet Problem XI, 1930, Report of the Commander in Chief United States Fleet*, Admiral W.V. Pratt, USN (14 July 1930), 53.

9. Admiral Frank H. Schofield, *United States Fleet Problem XIII, 1932, Report of the Commander-in-Chief United States Fleet* (23 May 1932), 11, 42.

10. At this time, the U.S. Navy's General Board set requirements for new ship designs. See Norman Friedman, *U.S. Destroyers: An Illustrated Design History* (Annapolis: Naval Institute Press, 1982), 86–90.

11. "Revision of Destroyer Tactical Instructions," Chief of Naval Operations, 3 December 1931, Naval War College Archives, Record Group 8, Box 40; *Tactical Employment Destroyers, U.S. Fleet, 1932*, 1 July 1932; *Entry 337:U.S. Navy and Related Operational, Tactical, and Instructional Publications*, Record Group 38, Records of the Office of the Chief of Naval Operations, NARA [hereinafter *Entry 337*], Box 67.

12. "Light Force Operation Plan No. 6–37," 1 April 1937, Annex Baker; "Fleet Problem XVIII—Comments., Commander Cruisers, Scouting Force," 3 June 1937.

13. David C. Evans and Mark R. Peattie, *Kaigun: Strategy, Tactics, and Technology in the Imperial Japanese Navy, 1887–1941* (Annapolis: Naval Institute Press, 1997), 276–80.

14. *Report of Gunnery Exercises, 1939–40*, Office of Naval Operations, Division of Fleet Training, 1941, *Entry 336A: U.S. Navy Technical Publications*. Record Group 38, Records of the Office of the Chief of Naval Operations, NARA [hereinafter *Entry 336A*], Box 59, 380.

15. "Change Number 4 to Cruisers, Scouting Force, Tactical Bulletin," 31 January 1940, *Entry 337*, Box 110.

16. *Night Search and Attack Operations, Destroyer Tactical Bulletin No. 5–38*, Commander Destroyers, Battle Force, 1938, *Entry 336A*, Box 129, 16.

17. "Night Division Battle Practice 1938–39—Destroyers, Battle Force, - Summarized Comments on," Commander Destroyers, Battle Force, 23 March 1939, *Fleet Training Division General Correspondence*, Record Group 38, Records of the Office of the Chief of Naval Operations, NARA [hereinafter *Training*], Box 224, 12; "Night Division Battle Practice, 1939–1940—Destroyer Flotilla ONE Summary." Commander Destroyer Flotilla One, 21 June 1940, *Training*, Box 276, 12.

18. Friedman, 99, 111–18.

19. Hone, *Learning War*, 323–41.

20. "Report of night action 11–12 October, 1942," Commander Task Group 64.2, 22 October 1942, *World War II Action and Operational Reports*. Record Group 38, Records of the Office of the Chief of Naval Operations, NARA [hereinafter *Action*], Box 17.

21. Operation Plan No. 1–42, Commander Task Force Sixty-Seven, 27 November 1942, *Action*, Box 241.

22. Vincent P. O'Hara, *The U.S. Navy Against the Axis: Surface Combat, 1941–1945* (Annapolis: Naval Institute Press, 2007), 130–37.

23. "Memorandum for Task Group Sixty-Four Point Two," Norman Scott, 9 October 1942, *Action*, Box 19.

24. "Report of Action," The Commanding Officer, U.S.S. Fletcher, 15 November 1942, *Action*, Box 984.

25. *Pacific Fleet Tactical Bulletin No. 4TB-42*, C.W. Nimitz, 26 November 1942, *Records of Naval Operating Forces*, entry 107, Record Group 313, NARA [hereinafter *Forces*], Box 6543.

26. "Outline of CIC Lectures for Destroyer PCO's and PXO's," COTCLant, 1 May 1944, *Training*, Box 1162; Hone, *Learning War*, 210–15.

27. *Pacific Fleet Tactical Bulletin No. 5-TB42*, C. W. Nimitz, 14 November 1942, *Forces*, Box 6543; *U.S. Pacific Fleet Tactical Bulletin No. 5–41*, W. S. Pye, 24 December 1941, *Forces*, Box 6543.

28. *Secret Information Bulletin No. 5: Battle Experience, Solomon Islands Actions, December 1942–January 1943*, United States Fleet, Headquarters of the Commander in Chief, 15 April 1943, Naval War College Archives, Record Group 334, Box 443, 31-2 through 31-6; Hone, *Learning War*, 163–207.

29. Operation Plan No. 2–42, Commander Task Force 67, 17 December 1942.

30. Operation Plan No. 6–43, Commander South Pacific Area, 27 February 1943.

31. "Action Report—Battle of Kula Gulf and Bombardment of Munda and Vila-Stanmore Areas, Night of 5–6 March, 1943," Commander Task Force Sixty-Eight, 9 March 1943, *Action*, Box 243; E. B. Potter, *Admiral Arleigh Burke* (Annapolis: Naval Institute Press, 1990), 72.

32. AR, Commander Task Force Sixty-Eight, 9 March 1943; Potter, 72.

33. O'Hara, 167; AR, Commander Task Force Sixty-Eight, 9 March 1943.

34. "Fire Control Plan; night action March 6, 1943," Gunnery Officer, U.S.S. Montpelier, 9 March 1943; AR, Commander Task Force Sixty-Eight, 9 March 1943.

35. "Action Report—Battle of Kula Gulf and Bombardment of Munda and Vila-Stanmore Areas, Night of 5–6 March, 1943," Commander South Pacific, 17 May 1943.

36. Operation Order, No. 10–43, Commander Task Group 36.1, 1 July 1943; O'Hara, 173.

37. "Action Report, Night bombardment of Vila-Stanmore and Bairoko Harbor, Kula Gulf, 4–5 July 1943," Commander Task Group Thirty-Six Point One, 30 July 1943; Samuel Eliot Morison, *History of the United States Naval Operations in World War II*, 15 vols. (Reprint: Boston: Little, Brown, 1984), 6:156–57.

38. O'Hara, 170–73.

39. *Secret Information Bulletin No. 10: Battle Experience, Naval Operations Solomon Islands Area, 30 June—12 July 1943*, United States Fleet, Headquarters of the Commander in Chief, 15 October 1943, *Forces*, Box 6398, 51–5 through 51–8.

40. Operation Order, No. 4–43, Commander Task Force 18, 16 March 1943; "Action Report—Night Engagement off Kula Gulf during night of 5–6 July 1943," Commander Task Group Thirty-Six Point One [CTG 36.1], 1 August 1943, *World War II War Diaries*, Record Group 38, NARA [hereinafter *War Diaries*], Box 203.

41. AR, CTG 36.1, 1 August 1943; O'Hara, 173–76.

42. AR, CTG 36.1, 1 August 1943; O'Hara, 176–77.

43. AR, CTG 36.1, 1 August 1943; O'Hara, 176–78.

44. Hone, *Learning War*, 220.

45. AR, CTG 36.1, 1 August 1943.

46. "Action Report—Night Engagement off Kolombangara during night of 12–13 July 1943," CTG 36.1, 3 August 1943, *War Diaries*, Box 203; Morison, 6:180–81; O'Hara, 180–82.

47. AR, CTG 36.1, 3 August 1943; Morison, 6:182–84; O'Hara, 182–83.

48. AR, CTG 36.1, 3 August 1943; Morison, 6:184; Eric Lacroix and Linton Wells II, *Japanese Cruisers of the Pacific War* (Annapolis: Naval Institute Press, 1997), 433.

49. AR, CTG 36.1, 3 August 1943; "H.M.N.Z.S. 'Leander'–Action Damage Reports," 9 September 1943; Morison, 6:186; O'Hara, 184–85.

50. AR, CTG 36.1, 3 August 1943; ADM 199/1331 [copy provided to author by Dr. Corbin Williamson]; Morison, 6:187–88; O'Hara, 185–86.

51. AR, CTG 36.1, 3 August 1943; Morison, 6:188–91; O'Hara, 185–87.

52. ADM 199/1331.

53. CTF 31 Serial 050814, August 1943; Morison, 6:212–13.

54. Quoted in Potter, 83.

55. Morison, 6:213.

56. "Action Report for Night of August 6–7, 1943—Battle of Vella Gulf," Commander Destroyer Division Twelve [ComDesDiv 12], 16 August 1943, *Action*, Box 628; "Vella Gulf Night Action of August 6–7, 1943—report of," The Commanding Officer, U.S.S. Craven, 8 August 1943, *Action*, Box 73; C. Raymond Calhoun, *Tin Can Sailor: Life Aboard USS Sterett, 1939–1945* (Annapolis: Naval Institute Press, 1993), 123; Morison, 6:215–16.

57. AR, ComDesDiv 12, 16 August 1943; "Translation of Captured Japanese Document: Japanese Navy Torpedo School Publication: Battle Lessons Learned in the Greater East Asia War (Torpedoes), Volume VI," JICPOA Item No. 5782, 11 April 1944; Morison, 6:218; Captain Tameichi Hara et al., *Japanese Destroyer Captain: Pearl Harbor, Guadalcanal, Midway—The Great Naval Battles Seen Through Japanese Eyes* (Annapolis: Naval Institute Press, 1967), 176–81; O'Hara, 190.

58. AR, ComDesDiv 12, 16 August 1943; Calhoun, 123; O'Hara, 191.

59. AR, ComDesDiv 12, 16 August 1943.

60. "Vella Gulf Night Action," The Commanding Officer, U.S.S. Craven, 8 August 1943.

61. Morison, 6:233–37; O'Hara, 193–98.

62. "Surface ship action for night of 2–3 October 1943 off Kolombangara—report of," The Commander Destroyer Division Forty-Two, 14 November 1943, *Action*, Box 33; Morison, 6:239–43; O'Hara, 198–200.

63. "Advance Preliminary Report of Battle Action—SELFRIDGE, O'BANNON and CHEVALIER with Enemy Forces off Sauka, Vella Lavella Night 6—7 October 1943," Commander Destroyer Squadron Four, 13 October 1943, *Action*, Box 33; Morison, 6:243–47.

64. "Advance Preliminary Report," Commander Destroyer Squadron Four, 13 October 1943; Morison, 6:248–49; O'Hara, 204–5.

65. Morison, 6:248–53; O'Hara, 205–6.

66. "Advance Preliminary Report of Battle Action—SELFRIDGE, O'BANNON and CHEVALIER with Enemy Forces off Sauka, Vella Lavella Night 6—7 October 1943," Commander Task Force Thirty-One, 19 November 1943, *Action*, Box 33.

67. "Advance Preliminary Report," Commander Task Force Thirty-One, 19 November 1943.

68. *CIC Handbook for Destroyers, Pacific Fleet*, Commander Destroyers, Pacific Fleet, 24 June 1943, *Records of the CNO HQ, COMINCH*, Record Group 38, Records of the Office of the

Chief of Naval Operations, NARA, Box 614; Timothy S. Wolters, *Information at Sea: Shipboard Command and Control in the U.S. Navy, from Mobile Bay to Okinawa* (Baltimore, MD: The Johns Hopkins University Press, 2013), 225; Hone, *Learning War*, 230.

69. Wolters, 210–11.

70. AR, Commander Task Group Thirty-Six Point One, 3 August 1943.

71. "Action Reports—Task Force Thirty-Nine covering operations for Empress Augusta Bay and Treasury Island Echelons—Period 31 October to 3 November, 1943," Commander Task Force Thirty-Nine [CTF 39], 3 November 1943, *War Diaries*, Box 204.

72. ARs, CTF 39, 3 November 1943.

73. ARs, CTF 39, 3 November 1943; Morison, 6:310–11.

74. ARs, CTF 39, 3 November 1943; "Report on Surface Engagement with Japanese Imperial Force Approximately Forty Miles West of Cape Torokina, Bougainville Island on 2 November, 1943," Commanding Officer, U.S.S. Columbia, 2 November 1943; "Action Report, U.S.S. Denver—The Night Engagement off Empress Augusta Bay, 1 November 1943," Commanding Officer, U.S.S. Denver, 4 November 1943.

75. O'Hara, 211–12; Morison, 6:312–16.

76. Morison, 6:317n10; ARs, CTF 39, 3 November 1943.

77. "Action Report of Night Engagement off Cape Moltke on the Night of November 1st—2nd, 1943," Commander Destroyer Squadron Twenty-Three [ComDesRon 23], 4 November 1943, *Action*, Box 606.

78. "Action Report of the Night Engagement off Cape St. George on the night of November 24th-25th, 1943," ComDesRon 23, 26 November 1943, *Action*, Box 606.

79. AR, ComDesRon 23, 26 November 1943; Ken Jones, *Destroyer Squadron 23: Combat Exploits of Arleigh Burke's Gallant Force* (Annapolis: Naval Institute Press, 1997), 247.

80. Morison, 6:352–55; O'Hara, 216–17.

81. AR, ComDesRon 23, 26 November 1943; "Report of Surface Action off Buka on the Night of 24–25 November, 1943," Commander, Destroyer Division Forty-Six [ComDesDiv 46], 25 November 1943, *Action*, Box 634; O'Hara, 217–18; Morison, 6:355–56.

82. AR, ComDesRon 23, 26 November 1943; AR, ComDesDiv 46; O'Hara, 218; Morison, 6:356.

83. AR, ComDesRon 23, 26 November 1943; O'Hara, 218–19; Morison, 6:356–57.

84. O'Hara, 218–19; Morison, 6:356–57.

85. "Comments on Battles off Empress Augusta Bay, November 1–2, 1943, and off Cape St. George, November 24025, 1943," President, Naval War College, 13 January 1944, *Action*, Box 752.

86. AR, ComDesRon 23, 26 November 1943.

87. *PAC-10, Current Tactical Orders and Doctrine*, U.S. Pacific Fleet, Commander-in-Chief, U.S. Pacific Fleet, June 1943, v.

88. "Preliminary Action Report for Battle of Surigao Strait, 25 October, 1944," Commander Cruiser Division Four, 2 November 1944, *Action*, Box 610.

89. Preliminary AR, Commander Cruiser Division Four, 2 November 1944.

90. "U.S.S. McGowan (DD678)—Action Report-Operation for Capture, Occupation, and Defense of Leyte, Philippine Islands, including the Battle of Surigao Strait," Commanding Officer, 5 November 1944.

91. Anthony P. Tully, *Battle of Surigao Strait* (Bloomington, IN: Indiana University Press, 2009), 24–28, 43–47, 124–25.

92. Commodore Richard W. Bates, *The Battle for Leyte Gulf, October 1944, Strategic and Tactical Analysis, Volume V: Battle of Surigao Strait, October 24th-25th* (U.S. Naval War College, 1958), 321, 352–60; O'Hara, 249.

93. Tully, 152–53; although other accounts suggest a single torpedo hit *Fusō*, Tully provides convincing evidence that two scored.

94. Bates 361–77; Tully, 176–79; O'Hara, 249–50.

95. Bates, 438–44.

96. "Action Report, Night Surface Engagement Surigao Straits, Leyte, PI, 24–25 October 1944, Commander Destroyer Squadron Twenty-Four," 30 October 1944, *Action*, Box 610; Bates, 425–37, 524–31.

97. Morison, 12:221.

98. Bates, 493–504, 511–16; Morison, 12:221, 224; O'Hara, 254–56; "Action Report, Battle of Surigao Strait of 24 October 1944," Commanding Officer, U.S.S. *Richard P. Leary*, 7 November 1944, *Action*, Box 257.

99. Myron J. Smith Jr., *The Mountain State Battleship: USS West Virginia* (Richwood, WV: West Virginia Press Club, 1981), 129.

100. "Action in Battle of Surigao Straits 25 October 1944, U.S.S. West Virginia—Report of," Commanding Officer, U.S.S. West Virginia, 1 November 1944, *Action*, Box 1508; *Pennsylvania* did not fire at all and *Mississippi* fired a single salvo. See Bates, 479–84; Morison, 12:224.

101. Tully, 217–18.

102. Bates, 546–49; Tully, 255–57; O'Hara, 257–59; Morison, 12:233–38.

103. Trent Hone, "Countering the Kamikaze," *Naval History Magazine* 34, no. 5 (2020).

104. "Action Report," Commander Destroyer Division One-Twenty, 6 December 1944, *Action*, Box 646.

105. "Action Report—2–3 December 1944," U.S.S. Allen M. Sumner, 4 December 1944, *Action*, Box 646.

106. AR, Commander Destroyer Division One-Twenty, 6 December 1944; O'Hara, 281–85; Morison, 12:371–72.

107. *USF-2, General Tactical Instructions, United States Fleets*, Navy Department, Office of the Chief of Naval Operations, April 1947, 13–10 through 13–18.

CHAPTER 7. CONTROLLING THE CHOPS

1. The author acknowledges the contribution of Vice Admiral Harry DeWolf, the Von Bechtolsheim family, Gil Lauzon and, particularly, Commodore Jan Drent for his translation and analysis of Kriegsmarine documents. The maps, courtesy of Chris Johnson and Robin Brass, are from an earlier study in Douglas M. McLean, ed., *Fighting at Sea: Naval Battles from the Ages of Sail and Steam* (Toronto: Robin Brass Studio, 2008).

2. Plymouth *Daily Sketch*, 29 April 1944. RCN history is devoid of dramatic quotes of the John Paul Jones variety; DeWolf's probably stand as the most pugnacious.

3. Deck log HMCS *Haida*, 26 April 1944 (author's possession).

4. Derek Howse, *Radar at Sea: The Royal Navy in World War II* (Annapolis: Naval Institute Press, 1993), 58.

5. ADM 199/430, Mountbatten, "Report of Action Between 5th Destroyer Flotilla and German Destroyers on Night of 28/29th November 1940," 5 December 1940. See also Adrian Smith, *Mountbatten: Apprentice War Lord* (London: Tauris, 2010), 130–35.

6. From Peter Dickens, *Narvik: Battles in the Fiords* (Annapolis: Naval Institute Press, 1997), 77.

7. Rear Adm. A. F. Pugsley, *Destroyer Man* (London: Kimber, 1957), 59; Mountbatten, "Report of Action."

8. ADM 199/430, Pugsley, Report of Action, 2 December 1940; M. J. Whitley, *German Destroyers of World War Two* (Annapolis: Naval Institute Press, 1991), 113–14.

9. Sir Reginald Bacon, *The Concise Story of the Dover Patrol* (London: Hutchinson and Co., 1932), 120; Keith Bird, *Erich Raeder: Admiral of the Third Reich* (Annapolis: Naval Institute Press, 2006), 55.

10. Dickens, 107.

11. Directorate of History and Heritage (DHH) *Kriegstagebuch* [War Diary] (*KTB*), *Karl Galster*, 29 November 1940; Whitley, 111–14.

12. ADM 199/430, Capt. C. Harcourt (DOD (H)) minute, 2 December 1940; Capt. A. J. Power (ACNS (H)) minute, 7 January 1941; Rear Adm. T. Phillips minute, undated.

13. ADM 199/430, C-in-C Western Approaches, "Report of Action Between 5th Destroyer Flotilla and German Destroyers," 5 December 1940; DTSD minute, January 1941; DCNS minute, 21 January 1941.

14. Pierre Hervieux, "The Elbing Class Torpedo Boats at War" in *Warship Volume X*, ed. Andrew Lambert (London: Conway Maritime Press, 1986), 95–102.

15. David Pritchard, *The Radar War: Germany's Pioneering Achievement, 1904–1945* (New York: Harper Collins, 1989), 190–203; ADM 220/204, Admiralty, "R. D. F. Bulletin," 29 May 1943; D. E. Graves, "German Navy Radar Equipment, 1934–1945," (1990), DHH, 2001/10.

16. Howse, 181.

17. ADM 220/204, Admiralty, "RDF Bulletin No. 6," 3 March 1944; ADM 199/1038, C-in-C Plymouth minute, 13 November 1943.

18. Arthur Hezlet, *The Electron and Seapower* (London: Davies, 1975), 264.

19. See Michael Whitby, "'Foolin' Around the French Coast': The Challenge of Operation Tunnel," in *Warship 2022*, ed. John Jordan (New York: Osprey, 2022).

20. ADM 199/532, Director of Operations Division (Home), 4 July 1944.

21. ADM 199/1038, Roger Hill, "Night Action with Enemy Destroyers," 22 October 1943.

22. Air Historical Branch (AHB), *The RAF in Maritime War*, vol. 4, DHH, 79/599.

23. *KTB 4.Torpedobootflotille* (*TBF*), 23 October 1943; Whitley, 147.

24. ADM 199/1038, Ralph Leatham, "Action with German Ships," 13 November 1943.

25. *KTB 4.TBF*, 6–23 October 1943; ADM 199/1038, ACNS(H) minute, 26 January 1943.

26. ADM 199/1038, Leatham, "Action with German Ships"; Admiralty to Leatham, 10 February 1944; Norman Friedman, *British Destroyers and Frigates: Second World War and After* (Annapolis: Naval Institute Press, 2012).

27. NARA, Admiralty War Diary, "Admiralty to C-in-C Home Fleet, 28 December 1943."

28. The RCN Tribals, the first of which was laid down in a British shipyard in September 1940, incorporated lessons from war experience, and were thus "improved" versions of their RN predecessors. Most significantly, the location of the 2-pounder "pom pom" was exchanged with the former searchlight position, allowing far superior arcs of fire for antiaircraft defense. They also had a wider beam to increase stability and were "Arcticized" to make them more suitable for operations in northern waters. See Michael Whitby, "Instruments of Security:

The RCN's Procurement of Tribal Class Destroyers, 1938–1941," *The Northern Mariner* (July 1992), 1–15. https://www.cnrs-scrn.org/northern_mariner/vol02/tnm_2_3_1–15.pdf

29. ADM 1/13326, C-in-C Plymouth, 26 January 1943.

30. Admiralty, "Action Information Organization in HM Ships–Outline of Policy," Confidential Admiralty Fleet Order 1455/43, 15 July 1943.

31. ADM 1/13326, "Minutes of a Meeting to discuss Action Information Organization in Destroyers," 8 June 1943; F. A. Kingsley, ed., *The Applications of Radar and Other Electronic Systems in the Royal Navy in World War 2* (London: Palgrave MacMillan, 1995), 61, 167; W. A. B. Douglas, Roger Sarty, and Michael Whitby, *A Blue Water Navy: The Official History of the RCN in Second World War* (St. Catharines, Canada: Vanwell, 2006), 308–10; Norman Friedman, ed., *British Naval Weapons of World War Two: The John Lambert Collection* (Barnsley, UK: Seaforth, 2019), 47–8. Friedman's reference to all RCN Tribals having full AIO applies only after their 1944 refits in Canada following their initial European tours.

32. John Watkins, "Destroyer Action, Ile de Batz, 9 June 1944," *Mariner's Mirror* (August 1992), 308.

33. DHH, Equipment Cards for HMCS *Haida* and *Huron*; DHH, 89/235, Admiralty, *Gunnery Review*, July 1945; DHH, 91/79, Admiralty, "*Guard Book of Fighting Experience*," December 1942.

34. DHH, 81/520, DeWolf, ROP, 20 January 1944, *Haida* 8000. For the prewar RCN see Michael Whitby, "In Defence of Home Waters: Doctrine and Training in the Canadian Navy During the 1930s," *Mariner's Mirror* 77, no. 2 (May 1991), 167–77.

35. DHH, 81/520, DeWolf, ROP, 7 April 1944, *Haida* 8000; ADM 199/263, Lees, AR, 2 May 1944, 8.

36. ADM 199/532, Norris, ROP, 27 February 1944; Tyrwhitt, ROP, 2 March 1944; Leatham, "Report on Tunnel, Night 1st/2nd March," 25 March 1944; DHH, *KTB*, Operations Division, German Naval Staff, 2 March 1944.

37. DeWolf interview with Michael Whitby, 20 August 1987.

38. *KTB 4.TBF*, 16–23 October 1943.

39. F. H. Hinsley, *British Intelligence in the Second World War* (London: HMSO, 1984), v3/1:280.

40. In advance of some Tunnels, Coastal Command aircraft made a radar sweep down the intended patrol route, primarily to provide cover for special intelligence.

41. Admiralty, *Battle Summary No. 31: Cruiser and Destroyer Actions, English Channel and Western Approaches* (1945), 9.

42. *KTB 4.TBF*; DHH, 81/520 *Athabaskan* 8000, "Evaluation of Actions," April 1944.

43. Lees, AR.

44. ADM 199/532, DeWolf, Report on Action, 27 April 1944.

45. *Battle Summary 31*, 12; ADM 199/532, DeWolf, "Report on Action," 29 April 1944.

46. See Michael Whitby, "The Case of the Phantom MTB and the Loss of HMCS *Athabaskan*," *Canadian Military History* (Autumn 2002), at https://scholars.wlu.ca/cmh/vol11/iss3/2/

47. ADM 199/263, DTSD minute, 8 June 1944.

48. ADM 199/22, Admiralty to Leatham, 21 September 1944.

49. ADM 1/14402, Jones, "Report of Action," 19 April 1943.

50. DeWolf interview.

51. John A. English, *Monty and the Canadian Army* (Toronto: University of Toronto Press, 2021), 18.

52. ADM 199/22, Jones to Admiralty, 4 November 1944. Emphasis is Jones.'

53. Basil Jones, "A Matter of Length and Breadth," *The Naval Review* (May 1950), 139.

54. See Michael Whitby, "Planning, Challenge and Execution: The Seaward Defense of the Assault Area off Normandy, 6–14 June 1944" (forthcoming Brécourt Academic).

55. Due to production difficulties surrounding the new, complex twin 5.9-inch forward turret designed for the Type 39As, the early ships were initially fitted with a single mount forward, reducing their main armament to four guns. When twin turrets were finally fitted throughout the class mid war, it was found that in medium to high seas its additional weight had a detrimental effect on the destroyer's seaworthiness and speed. Whitley, 64–66.

56. Whitley, 201–3.

57. *KTB T-24*, 29 April 1944; DHH, *KTB Befehlshaber Sicherung West* [Commander Security West], 3–4 May 1944.

58. ADM 223/195, Admiralty to Leatham, 1700B/6 June 1944.

59. Michael Simpson, ed., *The Cunningham Papers: Volume II* (Aldershot, England: Ashgate, 2006), 202.

60. AHB, *The RAF in Maritime War*, vol. 5, DHH, 79/599, 8–9.

61. ADM 223/195–196, Admiralty to Leatham, 1230B, 1405B, 1911B, 2310B 8 June 1944, and 0005B 9 June 1944.

62. ADM 199/1644, Jones, "Report of Night Action," 14 June 1944, 1; *KTB, 8.Zerstörerflotille Am 9.6.44*, DHH. The *KTB* includes analysis by Von Bechtolsheim and senior *Kriegsmarine* staff.

63. Basil Jones, *And So to Battle* (Privately Published, 1985), 15; Leatham, "Destroyer Action," 18 June 1944, 2; Jones, "Night Action."

64. Namiesniowski, "Report of Action," ud; Leatham, "Destroyer Action."

65. *KTB, 8.ZF.*

66. ADM 199/1644, Barnes, "Report of Action," 9 June 1944; *KTB ZH-1*, 8–9 June 1944.

67. DeWolf, "Report of Action," 9 June 1944; *KTB Z-24*, 9 June 1944, DHH.

68. *KTB T-24*, 9 June 1944; ADM 199/1644, Rayner, "Report of Action," 9 June 1944.

69. DeWolf, "Report of Action."

70. *KTB, 8.ZF*, 6; Jones.

71. DeWolf.

72. *KTB, 8.ZF.*

73. *KTB, 8.ZF.*

74. DeWolf interview with Whitby, 2 November 1992.

75. *KTB, Z-32.*

76. *KTB, 8.ZF*; DeWolf, "Report of Action."

77. DHH, Kreisch, *"Stellungnahme des FdZ."*

78. Robert W. Love and John Major, ed., *The Year of D-Day: The 1944 Diary of Admiral Sir Bertram Ramsay* (Hull, UK: University of Hull, 1994), 85.

79. Rear Adm. Ian Glennie, RN, Admiral (D) Home Fleet, to Hibbard; Vincent P. O'Hara, *The German Fleet at War* (Annapolis: Naval Institute Press, 2004), 220–32; Douglas et al., 302–17.

80. *Battle Summary No. 31*, 1–21; Jones, "A Matter of Length and Breadth," 13; Jones, *And So to Battle*, 113; Friedman, *British Destroyers and Frigates*; and Alexander Clarke, *Tribals, Battles and Darings: The Genesis of the Modern Destroyer* (Annapolis: Naval Institute Press, 2021).

CONCLUSION

1. E. B. Potter, *Admiral Arleigh Burke* (Annapolis: Naval Institute Press, 1990), 73.

2. Sadao Asada, *From Mahan to Pearl Harbor: The Imperial Japanese Navy and the United States* (Annapolis: Naval Institute Press, 2006), 266.

3. "Initial Report on the Operation to Capture the Marianas Islands," Commander Fifth Fleet, 13 July 1944; "Operations in Support of the Capture of the Marianas—Action Report of," Commander, Task Force Fifty-Eight, 11 September 1944.

4. Ernest J. King, fleet admiral, USN, and Walter Muir Whitehill, commander, USNR, *Fleet Admiral King: A Naval Record* (New York: W. W. Norton & Company, 1952), 563.

5. Capt. Wayne P. Hughes Jr., USN (Ret.), *Fleet Tactics: Theory and Practice* (Annapolis: Naval Institute Press, 1986), 34–35.

6. Adm. Sir Studholme Brownrigg, "Gunnery in the Royal Navy," Adm. Sir Reginald Bacon (Ed.), *Britain's Glorious Navy* (London: Odhams, 1943), 221.

7. Hughes, 34.

8. Trent Hone, *Learning War: The Evolution of Fighting Doctrine in the U.S. Navy, 1898–1945* (Annapolis: Naval Institute Press, 2018), 317–46.

9. Captain Hara Tameichi, with Fred Saito and Roger Pineau, *Japanese Destroyer Captain: Pearl Harbor, Guadalcanal, Midway–The Great Naval Battles as Seen Through Japanese Eyes* (Annapolis: Naval Institute Press, 2011), 148.

10. Quoted in Jon T. Sumida, *In Defence of Naval Supremacy: Finance, Technology, and British Naval Policy, 1889–1914* (New York: Routledge, 1993), 56.

APPENDIX

1. Involving warships larger than an MTB on each side. These statistics and those that follow are based on a database of surface actions maintained by the editors.

ABOUT THE CONTRIBUTORS

VINCENT P. O'HARA is a naval historian and the author, coauthor, or editor of sixteen books, most recently *Innovating Victory: Naval Technology in Three Wars* (with Leonard Heinz, Naval Institute Press, 2022). Mr. O'Hara has also written articles published in many magazines and journals, including *Naval War College Review, Naval History, Seaforth Naval Review, Warship, MHQ, Storia MILITARE, World War II*, and others. He was the Naval Institute Press author of the year for 2015. Mr. O'Hara lives in Chula Vista, California.

TRENT HONE is an award-winning naval historian and a leader in the application of complexity science to organizational design. He is the author of *Learning War: The Evolution of Fighting Doctrine in the U.S. Navy, 1898–1945* (Naval Institute Press, 2018), and *Mastering the Art of Command: Admiral Chester W. Nimitz and Victory in the Pacific* (Naval Institute Press, 2022). He coauthored *Battle Line: The United States Navy, 1919–1939* (Naval Institute Press, 2006 and 2021). Mr. Hone's articles have been awarded the U.S. Naval War College Edward S. Miller Prize and the Naval History and Heritage Command Ernest M. Eller Prize; he earned second place in the 2017 Chief of Naval Operations Naval History Essay Contest; and he was the Naval Institute Press author of the year for 2018.

STEPHEN MCLAUGHLIN is an independent historian with a focus on the Russian and Soviet navies in the late-nineteenth to mid-twentieth centuries. In collaboration with R. D. Layman he wrote *The Hybrid Warship: The Amalgamation of Big Guns and Aircraft* (Conway Maritime Press, 1991) and is the author of *Russian and Soviet Battleships* (Naval Institute Press, 2003, reissued 2021). His work has appeared regularly in the prestigious annual *Warship* since 1991. He has contributed chapters on the Soviet navy in World War II and on the Imperial Russian Navy in World War I for other edited works published by Naval Institute Press. Mr. McLaughlin is currently working on a book on Russian and Soviet cruisers.

LEONARD R. HEINZ is an independent naval historian and coauthor (with Vincent P. O'Hara) of *Clash of Fleets: Naval Battles of the Great War, 1914–18* (Naval Institute Press, 2017) and *Innovating Victory: Naval Technology in Three Wars* (Naval Institute Press, 2022). He is a retired attorney and the designer of a number of war games with an emphasis on tactical naval simulations. He holds a history degree from the University of Pennsylvania, where he focused on diplomatic history.

REAR ADMIRAL JAMES GOLDRICK (RET.), served in the Royal Australian Navy and on exchange with the British Royal Navy. He commanded HMAS *Cessnock* and *Sydney*, and held various higher commands, including the multinational maritime interception force in the Persian Gulf in 2002. He has authored many articles and chapters. His major works include *Before Jutland: The Naval War in Northern European Waters, August 1914–February 1915* and *After Jutland: The Naval War in Northern European Waters, June 1916–November 1918* (Naval Institute Press, 2015 and 2018). Rear Admiral Goldrick was the winner of the 2018 U.S. Navy's Chief of Naval Operations Naval History Essay Contest.

ENRICO CERNUSCHI is an independent naval historian based in Pavia, Italy. He has authored several dozen books and monographs and more than four hundred articles in Italian, English, and French. Mr. Cernuschi also has contributed chapters to works published by the Naval Institute Press. His major works include *"Ultra" La fine di un mito* (Mursia, 2014) and *Sea Power the Italian Way* (Ufficio Storia Marina Militare, 2019).

JONATHAN PARSHALL is an independent military historian and technology worker. He is coauthor of *Shattered Sword: The Untold Story of the Battle of Midway* (Potomac, 2005), as well as numerous articles. In 1995 he founded the longstanding *Imperial Japanese Navy Homepage* (www.combinedfleet.com). He is an adjunct lecturer for the U.S. Naval War College. He was *Naval History* magazine's 2012 coauthor of the year.

MICHAEL WHITBY is the senior naval historian at the Canadian Department of National Defence. He has published widely on World War II and Cold War naval operations and policy, including as coauthor of the two-volume official history of the Royal Canadian Navy in World War II, *No Higher Purpose* and *A Blue Water Navy* (Vanwell Publishing, 2002 and 2007).

INDEX